PRAISE FOR BRUCE FEILER AND
Learning to Bow

"[A] delightful, moving, and incisive account. . . . Feiler uses anecdotes, historical background, and keen personal observations to reveal and explain an extraordinary nation."
—*Philadelphia Inquirer*

"A hilarious and revealing book [that] marks the debut of a formidable talent."
—JAMES FALLOWS

"A refreshingly original look at Japan. . . . This book is a revelation."
—*Atlanta Journal-Constitution*

"This book should be required reading."
—*Japan Times*

"Anyone seeking a better understanding of modern-day Japan and of Japanese perceptions of American society will benefit immensely from reading this rich account."
—*Booklist*

"Extremely funny and informative . . . delivers a message that the world will have to learn."
—*Newark Star-Ledger*

"Bruce Feiler is a keen and thoughtful observer."
—*New York Times Book Review*

"Feiler, a superb narrator and storyteller with a gentle, ironic sense of humor, also possesses a potent intellect that at moments blazes forth, illuminating everything in its path."
—*Washington Post Book World*

ALSO BY BRUCE FEILER

Looking for Class:
Days and Nights at Oxford and Cambridge

Under the Big Top:
A Season with the Circus

Dreaming Out Loud:
Garth Brooks, Wynonna Judd, Wade Hayes,
and the Changing Face of Nashville

Walking the Bible:
A Journey by Land Through the Five Books of Moses

Abraham:
A Journey to the Heart of Three Faiths

LEARNING
TO BOW

Inside the Heart of Japan

Bruce S. Feiler

HARPER ⬤ PERENNIAL

NEW YORK ● LONDON ● TORONTO ● SYDNEY

A previous paperback edition of this book was published in 1992 by Houghton Mifflin. It is here reprinted by arrangement with the author.

HarperCollins books may be purchased for educational, business, or sales promotional use. For information please write: Special Markets Department, HarperCollins Publishers Inc., 10 East 53rd Street, New York, NY 10022.

First Perennial edition published 2004.

Designed by Lisa Diercks

Library of Congress Cataloging-in-Publication Data

Feiler, Bruce S.
    Learning to bow: inside the heart of Japan / Bruce Feiler.
        p. cm.
    Originally published: New York: Ticknor & Fields, c1991.
    Includes bibliographical references and index.
    ISBN 0-06-057720-7
        1. Education, Secondary—Japan. 2. Japan—Social life and customs—1945–. 3. Feiler, Bruce S. I. Title.

LA1316.F46 2004
373.52—dc22   2004040537

            09 RRD 20 19 18 17 16 15 14

For my parents,
Jane and Ed Feiler,
above and beyond the commas

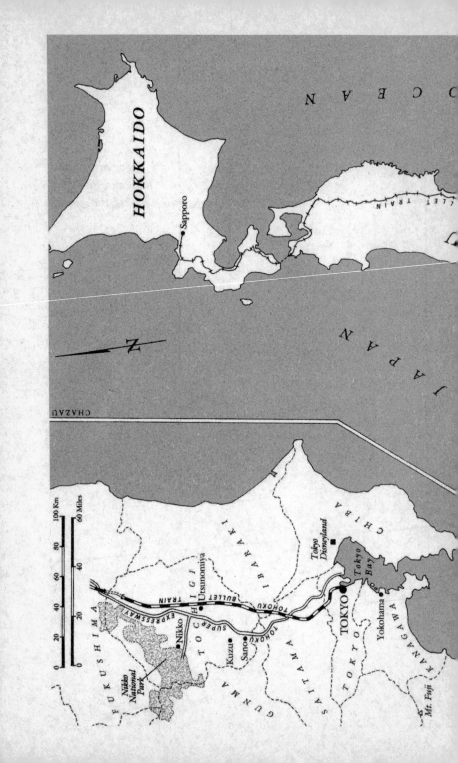

Oshieru wa manabu no nakaba nari.

Half of teaching is learning.

　—*A Japanese proverb*

CONTENTS

Learning to Bow

THROUGH THE OPEN DOOR

He drew a circle that shut me out —
Heretic, rebel, a thing to flout.
But Love and I had the wit to win:
We drew a circle that took him in.
                    — Edwin Markham, "Outwitted," 1915

I DROPPED MY PANTS and felt a rush of cool wind against my legs. Slower now, I slid off my remaining clothes to stand naked on the stone path, which felt warm below my feet. The smell of pine from the nearby hills lingered in the air. The sun had just set. It was a midsummer evening, my first night out of Tokyo, and standing bare on this mountain, I soon realized how quiet a body can be.

Unsure, I kept my eyes down, shifting first from my feet, now white with the chill, to my clothes, which lay in a shy heap on the grass, my pants still clinging to the shape of my body. Then suddenly I saw the other feet, and the legs. They too were bare. And as I watched them shuffling in my direction, my eyes told me what my mind had not time to know: these feet were looking at me.

Stepping back, I met the eyes that the feet belied and for a moment felt locked in a frozen stare. There were twenty-four eyes in all — open, agape, peering! — and despite all I had heard about Japanese eyes being narrow, these eyes seemed remarkably wide. As I stood on this mountain path, face to face with the twelve men who would be my hosts for a year as a teacher in their rural town, the only difference I noticed between them and me was that they were all wearing towels and I was not.

To my relief, one stepped forward. "Mr. Bruce," he said, offering a slight bow and a nervous laugh, "we are going to take a bath now. Perhaps you would like a towel."

I had never taken a towel into a bath or, for that matter, taken a bath with other people, but under the circumstances I agreed. "Thank you," I managed, trying to bow discreetly while drawing the small hand towel across my body.

As soon as I stretched it halfway across my waist, the others cheered, rushed forward, and with all the glee of a band of ten-year-olds parading a captured mouse, led me to the mouth of a nearby cave and the steaming, pungent fumes of a hot spring bath. As a newcomer in Japan, I would be welcomed into my office as I was welcomed into the world: with a bare body and a fresh bath.

Inside the cave, the bodies of other bathers emerged from the steam. They seemed to move slowly at first, as if muted by the weight of the thick white mist. Bare arms cut through the air, drawing handfuls of water to splash over shoulders; heads bobbed in the murky liquid like croutons in a gray broth. Some of the bathers — all men, I now realized — stood half sub-merged in the round pool, nodding their heads intently and speaking in echo; others floated quietly by, suspended by shad-ows of steam that lingered above the surface. From above, the pale evening light sifted through the air, giving the space the eerie feel of a Roman bath. But instead of wearing a toga, each of the men wandering outside the water held a small white towel over his private parts. As I watched these men clutching their towels while splashing and chatting and strolling about, I won-dered if I had discovered the secret reason behind bowing in Japan: to shake hands at a time like this and release the towel would mean a certain loss of face.

As we approached the water, the teacher who had earlier offered me the towel, a short, squat man with wiry black hair, a cherubic face, and a waddle that rocked him from side to side like a penguin, pushed the others away, put his arm around my shoulder, and led me forward.

"I . . . Mista Burusu boss," he said, tapping first himself and then me on the nose. "My namu . . . izu . . . Sakuragi. I amu Mista Cherry Blossom."

At this early stage in our relationship he spoke in English. Though my Japanese was far from fluent, I had the facility to understand most things when necessary and the ability to pretend not to when prudent. Both of these skills would prove vital to my survival.

Mr. Cherry Blossom led me to the wall of the cave and a row of men seated with their backs to the water. We sat on two round stones facing the wall and, with our towels draped over our knees, proceeded to douse ourselves in warm, chalky water from a shallow trough at our feet.

"In Japan," he said, this time in his native tongue, "we clean ourselves before entering the bath. Then we just soak in the water. This is our Japanese custom."

After pouring water over our shoulders with buckets and wiping our bodies with our hands (no soap), we were ready to step into the bath. Moving from the wall to where the water splashed at the edge of the pool, we walked slowly down the steps, slid up to our necks in the warm liquid, and for the first time removed the towels from below our waists and placed them — dripping wet — atop our heads.

"Doesn't it feel wonderful?" he said, closing his eyes, stretching his arms, and flashing a dreamy, self-satisfied grin.

"The water seems . . . alive," I said as I struggled to keep myself afloat while pushing away the flotilla of bugs swimming past my head. Then slowly my feet began to sink, and I realized that instead of being in a stone cavern, we were standing in an open mud basin with hot spring water bubbling up from the ground. With each burp from the earth, I would slide down further, until chalky liquid lapped at my mouth from below and dripped down my nose from above, where the wet towel slopped on my head. I closed my eyes and tried not to remember that I had just removed all of my clothes and washed myself with great care, all for the purpose of taking a bath in a giant bog of mud.

But I could not forget: "This is our Japanese custom."

I arrived in Japan in early August, at the time of year when the rusty orange afternoon sun lingers over hilltops for an extra hour at night, when the trees sit breathless throughout the day waiting for a whisper of evening breeze, and when people all across the land journey to the countryside for a brief summer repose.

"In Japan, we change with the seasons," Mr. Cherry Blossom explained. "What we eat, what we drink, what we say, all depend on the time of year." His dimpled face came alive as he told me tales of summer fruits and dancing fireflies. A former science teacher who now served as a regional curriculum adviser for the prefectural Board of Education, Mr. Cherry Blossom had an exaggerated bonhomie that reminded me of a bumbling chemistry teacher I had in high school who was nicknamed Mr. V. And so Mr. Sakuragi, whose name means cherry blossom, became Mr. C.

"When you write a letter to a girl," he continued, "you must always begin by referring to the season. 'It's summer, the air is hot, my heart longs for you.' And of course you should enclose a summer flower for her."

"What's a summer flower?" I asked.

"Red ones are best, like a hibiscus or a rose. Japanese girls all love them; they squeeze them on their lips."

Although I never saw a Japanese woman with hibiscus juice on her lips, I did come to appreciate the importance of these seasonal symbols: the song of the cicada after summer rain; colored leaves in autumn; snowdrifts in winter; cherry blossoms in early spring. The bath, an enduring and romantic symbol of leisure in Japan, is also steeped in the traditions of time. A bath in December, which relaxes the body after a strenuous day in the snow, differs from a bath in August, which legend says will "take the dampness from the body" after the rainy season. Mr. C told me proudly that his wife still places the skin of a citron in the family bath on the winter and the summer solstices. "In winter the citron keeps our bodies warm," he said. "In summer it cools us down."

Despite Japan's international status and its fascination with high technology, these cultural symbols from a fabled past remain alive in the collective imagination of modern Japan. A successful office, the Japanese insist, is one where the members build "a relationship without clothes on." In order to prime the caretakers who must tend the native spirit every day, the men of the Ansoku Education Office of the Tochigi Prefectural Board of Education took off three working days every August to welcome all new teachers, which this year included me, with a collective outdoor summer bath.

We had not soaked long in the water before my presence began to attract a crowd. Soon other men came wading over to our corner of the bath, sloshing through the mud and clutching their towels to their loins.

"This is Mr. Burusu," Mr. C said to the first well-wisher. He was stretching my name into Japanese form, in which all syllables end in vowels.

"This is Kato-*sensei*," he said, using the honorific *sensei*, which means teacher or master, instead of the more simple *san*. "He is my boss, but he is very fat."

Kato-*sensei* was a short, pudgy man with sagging cheeks, a bulbous nose, and wavy black hair that shook over his eyes as he hurried toward me past the other new teachers, gesticulating widely and leaving a sizable wake. I bowed slightly as he approached, being careful not to dislodge the towel from my head.

"Oh, oh, oh! So nice to meet you," he shouted, reaching out his hand and nearly embracing me. "You are a handsome boy." His greeting startled me; then I realized that he and others were admiring me not because I was particularly handsome but because I was white and tall. "I hear you are going to teach English in junior high schools," Kato-*sensei* continued. "I hope you teach Mr. Cherry Blossom, too. His English is very bad."

"You're crazy," Mr. C broke in. "My English is very good. Listen." He put his arm around my waist and straightened his back as if addressing a group of judges at a speech contest. "My namu Kazuo Sakuragi. I amu supaman. I amu hansomu boy."

The other teachers cheered and splashed water in support, but Kato-*sensei* pulled me toward him with a quick jerk. "No, no, no. Mr. Bruce is handsome boy. Mr. Sakuragi is crazy boy."

Suddenly our bath had become a cozy ménage, and I struggled to keep my head above water while these two middle-aged men clasped their arms around my back and compared me to a summer's day. Our party naturally attracted others, and soon most of the veteran teachers were sizing up the new man in town.

"He sure is tall," said one man.

"And his nose is high, too," observed another.

"He looks like a model."

Under these circumstances, I realized that developing a relationship without clothes on meant them appraising my physique.

"Does he have a girlfriend?" someone asked.

Just as I approached the point of total desperation, a group of young teachers appeared at the outer edge of the circle. They too wore towels, but they were much quieter than their superiors. Mr. Cherry Blossom hushed the crowd and gestured for the three to come forward. "New teacher, new teacher," he said, pointing back and forth from them to me in great excitement. They did not look very enthusiastic about cutting into this dance, but the shortest of the three, perhaps sensing my helplessness, stepped away from his friends and moved closer.

There, standing alone in the middle of two dozen bathers all pondering this low-tech import from abroad, he very calmly reached his hand out and in near-perfect English said, "Hello, my name is Cho Takashi. You can call me Cho."

A friendship was born.

I came to Japan at the invitation of the Japanese Ministry of Education, to teach English language and American culture in Japanese schools as part of a program to bring native English speakers into the heart of Japan. The scheme is part of a broad effort by the government to achieve what the Japanese call "internationalization," the buzz word for a new, more globally powerful, more globally conscious Japan.

As I traveled through rural towns, going into schools and homes and participating in the lives of Japanese people who might never feel the trickle-down effects of their country's newfound wealth, I came to appreciate both the difficulty and the necessity of internationalization. In my city of Sano, I was the first foreigner many of the people had ever seen in person. I was the first person they had met who had white skin, brown hair, and a "high" nose — one that sticks out from the face, not one that starts high on the forehead. I was the first person they had known who was not Japanese. Even among teachers, those trusted with telling the next generation about the outside world, I was an anomaly.

After the bath I stood again alongside my colleagues, shivering as steam floated off my warm body and bemoaning the fact that my towel was wet. Soon Mr. C brought me a folded white robe with a dark blue bamboo print.

"This is a *yukata*," he said, "a summer kimono. Please wear this to the party tonight."

I slipped the starched cotton over my shoulders and drew

the sash around my waist. The cloth clung to my wet body like papier-mâché to a balloon, except for the flaps in front, which dangled helplessly just above my knees. Feeling a bit exposed, I quietly pulled a pair of shorts up under my robe, but still I felt — and looked — like a half-wrapped candy bar.

"Let's have a party," the men shouted, grabbing my clothes along with theirs and pushing me down the path toward the lodge.

As we approached the three-tiered stone pavilion where we would be staying for the night, Mr. Cherry Blossom again eased close to me and put his arm around my shoulders.

"We want you to give a short speech tonight."

"A speech?" I said, glancing down at my bare knees.

"A self-introduction."

Before I could protest, we had arrived at the hall and entered a large banquet room with latticed beams on the ceiling, straw mats on the floor, and long rows of low tables lined with over 140 freshman teachers — now men *and* women — sitting on the floor with bended knees and decked out in identical white robes with the dark blue bamboo print. Were these teachers, I wondered, or dolls?

Mr. Cherry Blossom ushered me to the front of the room and a seat on my knees next to the "fat man" from the bath.

Beginning with Kato-*sensei*, each of the men at the table addressed the gathering, laboring through prepared speeches in which the teachers were admonished to work hard, to understand their role in society, and to do their best. Throughout, the recruits sat stone-faced and silent.

Finally the oldest man at the table raised a toast. The mood lightened, the beer flowed, and all the teachers shifted off their knees and settled onto the floor. After a half hour of eating sliced fish and shrimp and downing beer and whiskey, Kato-*sensei* again rose to his feet.

"This year we are having an international party," he announced. "We have an honored guest with us tonight, who has come from America to be a teacher in our prefecture. As you can see, he is a very handsome boy."

As I rose to stand beside him, the general murmur of the party dimmed and everyone turned to hear me speak.

"Good evening," I said, using my best speech-contest Japanese.

"Good evening," the crowd said in unison, mumbling to one another about how this foreigner could speak Japanese, as if I had just recited an original haiku.

"My name is Bruce . . .

"I come from Georgia, in the United States of America . . .

"This year I am going to teach my language and my culture to your students. I hope you will also teach me about Japanese culture." I mentioned something about the weather and then about the approaching school year. To end, I tried to add a bit of self-deprecation to my speech, as I had been advised by Japanese friends to do. "Since you and I are new teachers," I said, "I hope we can be friends. But my Japanese is very bad, so please speak English with me." Instead of solemn admiration, this last line brought unexpected laughter from the crowd, and I realized I had a long way to go before I mastered Japanese humility.

After thanking them for their patience and bidding them good night, I kneeled down as fast as I could without flashing open my robe. But the audience was giving me a standing ovation and shouting for an encore. Several teachers from the front row conferred with Mr. Cherry Blossom, then scampered over to my place and asked, "Won't you answer some questions?" Within seconds I had a microphone in my hand and Mr. C was laughing merrily at my side, telling the troops they were free to ask whatever they wanted. It was open forum on the foreigner: seven score of drunken Japanese teachers versus one very sober American man.

The first questions came slowly.

"Can you eat Japanese sushi?"

"Can you drink Japanese sake?"

"Can you use Japanese chopsticks?"

They came in Japanese in rapid succession, first from one side of the room, then the other, and I answered them in English, assuring the crowd that, yes, I could eat Japanese food and drink Japanese wine. Then, as everyone became a little more relaxed, the questions took a different tack.

"Do you like Japanese girls?"

"Do you want a Japanese girlfriend?"

"Who is the prettiest girl in this room?"

These questions came faster and faster and I bobbed through the blizzard, feeling a bit like an amateur host of *Dr. Ruth Does Japan*, when suddenly a question in English came floating from the back of the room.

"Do you like sex?"

I was stunned. Surely I had misheard. But no relief came, and I stood still on the mat, silence creeping across the room, ten inches of my leg showing between my robe and the floor, a microphone in my hand. Mr. C glanced at me with a puzzled look on his face, and I realized that he had not understood. Taking this as a cue, I turned back toward the audience, smiled, and said, "Yes. I like sushi very much."

Every room I entered in Japan I entered through the same door — one that led from the outside in. Every time I entered a home, a school, or a classroom, I was treated as if I had just walked through that door for the first time. To those inside, my world seemed exotic and far away, so they assumed their world was strange and exotic to me. When I first went to Japan during the previous year to study for six months at a university, I knew I would be an outsider, an American in Japan. But when I returned to teach in Tochigi, I imagined that the longer I stayed in the country, the more I would be welcomed into every room as if I were an insider.

Others who have written about living in a foreign land have described the shifting moods of affection and disaffection the foreigner feels — one day enamored with the host culture, privileged with the secret access to the heart of another world; the next day dismayed at always being kept away from its inviolable core. I too felt these swings in emotion, yet those around me had no idea that my feelings about Japan were evolving. To them I remained a newcomer, and long after they ceased being exotic to me, I remained exotic to them.

Several weeks into my stay, Mrs. Cherry Blossom, a plump, jolly woman who taught home economics to junior high school girls and raised two junior high school boys of her own, hosted a welcome party for me along with some of her friends. As the party began, spread across the table at the center of her living room floor was a marvelous assortment of traditional Japanese party fare: heaping trays of sushi; bowls of pickled vegetables, tofu, and potatoes; plates of salads and compotes. In front of me, however, she had discreetly placed a small plate of egg salad sandwiches with a knife and fork tucked beneath a napkin. Moved by her thoughtfulness but by then quite accustomed to dining with chopsticks, I plucked a pair from the center of the table and joined with the other guests in prying nuggets of food from the trays and putting them into my mouth.

I had not lifted the first bite of raw fish halfway to my lips when the whole conversation stopped dead and everyone turned to marvel at my unimaginable skill.

"That's amazing," swooned the lady to my left as she focused her glasses on my fingers.

"So skillful," said another.

They were so generally impressed that they beckoned our hostess from the kitchen to witness this display of manual dexterity by the foreigner. She came rushing to my end of the table, dripping her serving spoon into my lap, and exclaimed, "Can you use chopsticks?"

I heard this type of question nearly every day I lived in

Japan. Can you drink Japanese beer? the teachers asked me. Aren't you afraid of Japanese thunder? And, in a devastating irony that was repeated every time I took a drive with someone, Can you fit into a Japanese car? Most people assumed that no matter how hard I tried, no matter how long I lived in Japan, I could never tolerate Japanese customs. The longer I stayed in Japan, the farther I moved into the room and away from the door, yet I never escaped the penumbra of that door frame, which seemed to follow me wherever I went.

Over time, I came to feel that the problem lay in a simple misconception: most Japanese believe that only they can understand Japan. Near the end of my stay, a time of great political scandal in the nation, I was the guest of a seventy-five-year-old retired English teacher and political activist, Gunji-*sensei*.

"Japanese politicians never speak straight," he said with great conviction as we sipped tea and ate rice wafers one Saturday afternoon. "Just the other day a member of parliament appeared on television, mentioned something about an illness, and announced that he would retire. He didn't say that he stole money. He didn't say he was corrupt. He said only that he was 'ill.' I bet you can't catch the meaning of what Japanese people say. You don't understand how they think or how they speak."

I reminded him that I had been a *sensei* in Japanese schools. I had stood alongside teachers in classrooms. I had eaten school lunch with students and joined in sports days and school excursions, where Japanese boys and girls learn how to behave in a group and how to speak indirectly.

"But we are one race," he insisted. "We are unique. Only a Japanese person can understand the heart of another. You can't figure us out because you are a foreigner."

"That isn't true," I maintained. "It is only a myth that Japanese children learn in school. Japanese can understand their politicians because they know how to interpret their words. After being in school in Japan I can do the same. Sometimes I even think like a Japanese myself. It's not magic, it's government policy."

The widespread myth of Japanese uniqueness — part old wives' tale, part cultural obsession — plays two ways. The pleasant surprise that people proclaim when they discover a Westerner using chopsticks often becomes arrogance when they insist that Japanese rice — or beef, or beer — is better than that in other countries. This issue of uniqueness has become the number one cultural dilemma facing Japan today. Some argue that the country should overthrow its legacy of isolation and speed up its integration into the rest of the world, while others believe Japan should resist the influx of Western values and stress its own distinctive heritage. The chief battleground in this debate lies in public school classrooms and in the minds of the next generation. Should Japanese schoolchildren eat bread or rice for school lunch? Should they eat with a fork or with chopsticks? Should textbooks contain less or more information about Japan's militaristic past? Should students be allowed to wear Mickey Mouse emblems on their socks at school?

While the means may be debated, the primary goal of Japanese schools remains essentially unchanged: to produce good citizens, those who are committed to thinking beyond themselves and to advancing the needs of the country. From the opening day of elementary school to graduation day from junior high, students hear of the opportunities and obligations of being members of the Japanese nation. To understand Japan — its work ethic and its strong identity — one must understand these lessons as they are taught in schools. By the time students have finished ninth grade, the end of their compulsory education, most understand the sacrifices they must make to fit into society and are willing and able to make them. Most of these students will enter the system and become, in time, other well-qualified cogs, "Made in Japan."

· · · ·

At the end of the party, the new teachers disbanded into smaller groups, dispersed into various rooms in the lodge, and gathered around whiskey bottles and sushi plates for less formal initiation parties. Mr. C, now scarlet from the flush of the beer, led me from room to room, to a myriad of new chances to test my resistance to liquor and my ability to perform under stress.

Each group had new topics to discuss — "Do all Americans smoke marijuana?" — and new stunts to perform — "Can you drink an entire shot of whiskey in one gulp?" After several hours of this inaugural protocol, Mr. C retired to his room and left me in the hands of one last group of teachers, who were determined to see me dance. They had in mind the perfect stunt for a new teacher — to sing the most famous children's song in all Japan, "Mount Fuji." And so it came to pass that at two-thirty in the morning in the mountains of central Japan, on the third floor of a lodge closed for the night, I stood in my bare legs and bathrobe, waved my arms, danced, and sang the song that Japanese youth have sung for a thousand years:

> *With its head held up so high,*
>   *way above the clouds,*
> *Looking down on other mountains,*
>   *'round on all four sides,*
> *Listening to the god of thunder,*
>   *way above the land,*
> *Fuji is the number one . . .*
>   *mountain in Japan.*

I ended the song with my hands clasped above my head in a triangle that vaguely resembled Mount Fuji, and I suddenly felt like the mountain itself, "looking down on other teachers, 'round on all four sides." I thanked those left standing, bowed, and made my way into the hall toward the room I shared with Mr. C and the other teachers from my office. Arriving at the door, I saw that bedrolls had been laid across the padded floor and the men from my office lay sprawled amid twisted sheets and pillows like teenagers at a slumber party after a game of strip poker. I tiptoed through the mass of bodies to the last empty roll, against the far screen wall. Slipping off my robe, I slid between the sheets with a sense of great anticipation, until, to my great despair, my feet came popping out the other end of the bedroll. Realizing that nothing would cover me tonight, I lay on my back, closed my eyes, and in a moment of peace, dreamed that I was in Japan — surrounded by teachers, wearing no clothes.

2

RED LIGHTS AND GREEN TEA: DRAWING THE LINES

Each ancestor, while travelling through the country, scattered a trail of words and musical notes along the line of his footprints, and these lay over the land as "ways" of communication between the most far-flung tribes.

— *Bruce Chatwin,* The Songlines

THE TOHOKU SUPER EXPRESSWAY stretches eight lanes wide and reaches over eight hundred kilometers north from Tokyo into the hinterlands of central Japan. From its start near the heart of the city, the road winds its way through the bustling urban sprawl of the capital, spans two colorless rivers hurrying toward their dénouement in the depths of Tokyo Bay, then heads inland toward the vast provincial region that the Japanese call *inaka*, or what people from my home in Georgia would refer to as "up the country." Soon the clamor of neon fades from view and steel girders cease piercing the sky. The endless queues of cramped apartment blocks and towering concrete factory walls slowly back away from the advancing road as if to make room for the tile-roof buildings and moistened rice fields that will soon take their place. Several hours down the road, a small city rises from the grassy lowlands where the Tohoku bends toward the north. In the parking lot just off the highway, a towering sign salutes the new arrival: "WELCOME TO SANO. PATIENCE AND HUMILITY PREVENT ACCIDENTS."

Sano is a city — a *shi*. In Japanese, places are separated into categories by size and then labeled with identifying tags, not unlike socks in a well-ordered drawer. The classification is based on population — up to ten thousand is a village, ten to twenty thousand is a town, and above that is a city. Thus the city of Sano becomes Sano-*shi*, and the town of Kuzu, Kuzu-*machi*.

For the newcomer such names are helpful. That is especially true in this community, where one might never guess from its surface that Sano — whose two Japanese characters translate somewhat forebodingly into English as "left field" — could actually qualify as a city. Although it has a population of close to fifty thousand people, one train station, two department stores, and more than its share of stoplights, Sano has no curbs. Downtown, buildings seem to melt into streets, and residents step out of their homes in the morning into lines of oncoming traffic. In summer, when people in cars roll their windows down and people at home leave their doors ajar, a driver can sit in his car at a red light and watch a baseball game on a TV set in a house by the side of the road.

Leaving my two-room second-floor apartment each morning, I would stop to admire the delicate Japanese garden that my landlady had tended for over sixty years. Short-leaf pines and persimmon trees hunched over the tiny courtyard, a pale green moss shrouded the ground, and three aging carp mingled quietly in the still, black waters of a homemade pond. Retrieving my bike from astride her stone fence, I would pedal down the narrow lane, past the barbershop and the lean stone pillars of our neighborhood Shinto shrine, and head toward the shadow of the Hotel Sunroute, an eight-story beige concrete box that was the tallest and ugliest building in Sano. As I crossed the railroad tracks every day heading away from town, the same women would be gathered outside the same stores, having, it seemed, the same conversations — "Sure is hot today." "How is your daughter doing?" "Did you hear the mayor is having an affair?" These women would bring out their burnable garbage for collection on Tuesdays, their bottles on Wednesdays, and their used batteries in small plastic bags every other Saturday morning. Most of them had probably never been to the Kentucky Fried Chicken on the other side of town or the Mos Burger Store just around the block.

Before coming to Japan, I had often heard that Japan is the wealthiest country in the world. I had read stories about toilets that talk and robots that answer the telephone. I had seen pictures of fancy buildings all over the world which the Japanese recently bought. With this introduction I half expected to find an island paradise overflowing with expensive cars, spiral escalators, and extravagant buildings that the Japanese already owned. But in Sano I found a world quite different from the polish and poshness of Tokyo and far closer to the disheveled tin-roof towns I remembered from my childhood in the American South. Although Japan has the highest per capita gross national product in the world, the lives and homes of most Japanese people do not reflect this statistic. My apartment, for example, had no heating, no insulation, no hot running water in the sink, and no overhead lighting. My toilet had no seat. Still, my Japanese friends told me I had the nicest apartment they had seen in town. My American friends, meanwhile, had a different name for this lifestyle: they called it creative camping.

After a ten-minute ride through the traffic, I would arrive on the outskirts of town. Here, away from the crowded downtown alleys and narrow single-lane streets, away from the faded plastic hydrangeas that drape the main street in summer, away from the gray tin aura of old-town Sano, a shining new building appears along the road. The five-story hall is wrapped in white-washed stucco and gray metallic windows. Along with the other trophy towers in town, the Hotel Sunroute, the Jusco Department Store, and the "Happy Home" Wedding Palace, it has one of the few elevators in Sano. On the top floor of this government building, the elevator doors open directly across from the Ansoku Regional Branch of the Tochigi Prefectural Board of the Japanese Ministry of Education.

It was here, on a Monday morning in August, fresh from my inaugural bath, that I first reported for work.

According to legend, when the American army moved into Japan at the end of World War II, they brought their own desks — large, solid, U.S. government–issue, gray metal desks. Perhaps because of their durability, perhaps because of the Japanese custom of adopting things foreign, perhaps because of mere fashion, these desks have remained ever since. To this day, these indestructible desks are the staple of government, and many non-government, offices across Japan. On my first day as a government employee in Japan, I too was given my very own gray metal desk, with a hard-back gray metal chair to match.

"This is Mr. Bruce's desk," Mr. Cherry Blossom said as I arrived, using his junior high school English in front of his colleagues and plopping down a sign that said, indeed, "MR. BRUCE'S DESK."

The crowded room had three groupings of desks spaced evenly across the white tile floor. My desk was in the middle section of nine, along with Mr. C's. Our group was arranged in tight formation, with four desks lined up end to end like football players in a line of scrimmage, facing four others directly across. The section chief, like a referee, was perpendicular at the top. Since all the desks in this double-file line were touching, they formed what amounted to a giant tabletop that stretched from the door to the plate-glass windows overlooking downtown Sano. In this huddle, every conversation and every minor memo became the business of the whole group. Every four desks, moreover, shared one telephone.

As soon as Mr. C had introduced me to my seat, two women came rushing over with hot rags to wipe off the top of the desk.

"This is Arai-*san*," he said, pointing to the older of the two secretaries, whom the Japanese call "O.L.s," office ladies. She was a middle-aged woman with a tapered neck, an elongated face, and black hair pulled back from her face.

"How do you do," she whispered shyly, drawing her hand to her mouth and grabbing her assistant for protection. The younger woman, in her mid-twenties, was taller, with straight hair that hung down her back and a timid gaze across her face like that of an animal frightened from sleep.

"This is Eh-*chan*," Mr. C said, resisting what seemed like the temptation to pat her on the rear. "She is not yet married."

Slowly, allowing for bows and handshakes, Mr. C led me around the phalanx of desks and formally introduced me to each person.

"This is Mimura-*sensei*. He teaches science.

"Mogi-*shidō*. He is a student adviser.

"Nanmoku-*kachō*. He is the section head."

At first glance all the desks in this formation appeared equal, like beds in a hospital ward, but a closer look revealed a subtle order. Younger people sat at the bottom of each line, seniors higher along, and the section chief alone at the end. As a worker advanced along this route, he was allowed to keep more paper on his desk, more cushions on his chair, and perhaps even a pair of slippers underneath. In this office, one desk stood out. It was placed in the center of the room, separate and unequal from the rest, and its chair was draped in a sheepskin rug. Above the desk a lone sign dangled from a fluorescent light. It read, simply, *shochō*: Director.

"Welcome to our office," said Kato-*sensei*, now tucked into a blue polyester suit and looking less fat than he had in the bath. "Please have a seat." He gestured toward a brown vinyl sofa and two easy chairs that were grouped together in front of his desk. Arai-*san* went dashing across the hall to a small kitchen and returned with two cups of tea and a basket of crackers, which she set on the knee-high coffee table between the sofa and the chairs.

"Can you eat Japanese crackers?" she asked. "They are made of Japanese rice."

I assured her that I could.

"Ahhhh," she said, nodding her head up and down and resting her finger on her chin. "That's amazing."

Kato-*sensei* and I chatted for a while. He asked about my new apartment and offered to buy me a rice cooker. Then he started to explain my duties. This would be my home office, he said, but after the start of the fall term the following week, I would come here only on Wednesdays and visit schools on all other days. Most of the time I would be teaching at Sano Junior High, although I would have to make short visits to many of the other schools in the area. Unlike Japanese teachers, I would have Saturdays off. "If you have a problem," he said, "you can talk with Mr. Manager, Mr. Personnel Director, or Mr. Section Chief. They will be glad to help you."

In addition to reading desks in Japan, one must also be deft at hearing titles. All the men in the office were former teachers and were thus entitled to have the word *sensei*, teacher, attached to their names. But in this office, *sensei* was just the minimum. Mr. C was also a *shidō*, adviser; Nanmoku was a *kachō*, chief. When they addressed one another, the teachers dropped their names and used titles instead: "Mr. Section Chief, telephone for you."

Even the titles for the two secretaries showed rank. The older woman, Arai, earned the term *san*, and the younger only *chan*, a diminutive term used mostly for children and unmarried girls. Like the difference between a *shi* and a *machi*, the difference between a *san* and a *chan* is part of a cultural code that maintains order and assures that hierarchy is preserved.

For most of my first days in Sano, I participated in introductions of this kind, known formally as *aisatsu*. In less than five days I was introduced to all the other government workers in my building, the mayor, the head librarian, the director of the Public Health Department, the manager of the train station, and — to cover all bets — the chief of police. At the end of my first week, Mr. C decided I was ready to meet my most important protector, the principal of my school.

Sano Junior High School occupies a small plot of land on the west side of the city, at the foot of a large span of rice fields leading into the mountains of central Tochigi. Although the school was founded just after the Second World War, the building itself is new, standing three stories tall, with the same white-washed stucco as the government office building across town. The front of the school faces an open parking lot and a covered courtyard where students leave their bicycles, and the back of the school overlooks an enclosed dirt field, bordered by two soccer goals, a swimming pool, and a gymnasium. With its thick walls, lack of landscaping, and concrete, sterile air, the building looked to me more like a prison than a schoolhouse.

Once we were inside, Mr. C led me into a big, open room where several dozen people were working. Reading the desks, he quickly discerned the most prominent person in the room and asked him if the principal was in his office. Within seconds a large, broad-shouldered man dressed in a light gray suit, white shirt, and dark blue tie shuffled out to meet us. His face was square, with a flat, tense smile and thick black-rimmed glasses. His glistening gray hair had been stretched taut with a comb and matted to his head with a brand of hair tonic that smelled like distilled vinegar. Mr. C apologized for the inconvenience, bowed, and asked for permission to perform an introduction. Without objection, the principal excused himself before his colleagues and announced that a formal greeting was in order. All those at their desks dutifully rose and pulled on their coats from the backs of their chairs.

With everyone in place, the principal addressed the office

and introduced the go-between, Mr. C. We bowed. The go-between then introduced the guest of honor. Again a bow. I then introduced myself. Another bow, this time deeper. Finally the principal thanked all of us for our consideration, and life returned to normal. We bowed a final time.

At this point we were shuffled into the host's receiving room and invited to partake in a customary snack. The principal's room was spacious and carpeted but had the same gray metal desk and brown vinyl furniture as every other office I had visited.

"Please sit anywhere," he said to me.

When I first began my round of visits I accepted offers like this at face value and sat in any seat. After a while, however, Mr. C pulled me aside and told me that I should not sit just anywhere but should confine myself to the couch. The easy chairs are for the host, he said, and the couch is for the guests.

"*Hajimemashite*," the principal said. "We are meeting for the first time. My name is Sakamoto." He held his business card with two hands above his head and bowed until both his head and the card dipped below his waist. I rose to accept his offering and bowed deeply in return.

"I am the principal of this school," he said. "Do you see those pictures on the wall?" He pointed to a series of tinted black and white photographs that hung against one of the paneled walls. "There have been twelve principals before me in the forty-year history of this school. You are the first foreign teacher ever to come here. We are honored to have you with us."

Soon an office lady appeared with three cups of green tea, three lacquer saucers, and a plate of sponge cakes, which were stuffed with purple bean paste, wrapped in plastic, taped, and placed in a box with a ribbon. For several minutes no one acknowledged the tea; then the principal gestured and said, "Please," and we partook with a short apology and an expression of thanks.

As the guest, I remained quiet through most of these meetings while the go-between spoke on my behalf. But this host was eager to speak with me.

"I hear you come from Georgia," he said to me. "I like *Gone With the Wind* very much."

Mr. C nodded and looked at me in surprise. He hadn't known that this movie took place in my home state, he said.

"Do you know what that is?" the principal asked, pointing to a small flowerpot on the floor. A bare trunk stuck out of the dirt and several naked branches protruded from the side.

I told him I didn't know.

"It's a cotton bush!" he cried. "Just like you have in Georgia."

"I didn't know cotton grows in Japan," I said.

"It doesn't," he said. "We don't have the right soil. Plus we don't have any slaves . . ." He paused as if to consider his next line. Then slowly a smile crept across his face. "All we have is our wives." The two men laughed uproariously at this comment: Mr. C, a wiry little pug with a chipper laugh; and Sakamoto-*sensei*, a Great Dane with a sturdy bass guffaw.

I assured them we didn't have slaves anymore either, but they didn't seem to listen. After a moment Mr. C jumped up from the sofa and pulled me up beside him.

"You see, Mr. Bruce, I'm sure Mr. Principal will take good care of you. He already knows your heart."

Aisatsu greeting ceremonies like this occurred in school and in my office no fewer than eight times in the course of a normal day — and more during busy seasons. That meant eight times a day everyone must rise and bow to the guest. Eight trays of tea that must be made, served, drunk, retrieved, and cleaned. Eight plates of sponge cakes stuffed with bean paste, fancy rice crackers, cookies, or plastic containers of jelly that must be eaten. It didn't seem to matter what was said in these meetings. Their purpose was to establish the unofficial paths along which tacit deals and arrangements are made. As I was making my way around Sano, drinking tea and bowing to strangers, I was creating lines along which I could later walk, if need be. In Australia, the Aborigines called such paths "songlines." In Japan, they are called *ningen-kankei*, the web of human relations.

"Japanese culture runs on *ningen-kankei*," Mr. C explained to me after our visits were done. "We Japanese like to work with people we know. Japan has one race, so we never had strong laws. The laws of human relations are our laws."

My initiation into Japanese laws was swift. Early in my stay, I developed the habit of speeding through town on my bicycle with some abandon. Darting in and out of streets, at times I startled local children, and at other times a storekeeper flagged me down in order to ask one of those questions — "What are doghouses like in America?" If during one of these journeys I happened to come to a red light, and if the hour was late and the roads were empty, I would ride through that red light and continue on my way. That was my own law, but the law of human relations worked differently.

Sure enough, some citizen in town eventually witnessed my showing such wanton disrespect for the law, and she telephoned my office to express her disapproval. Like an insect flying unaware through the woods, I had been trapped in a web. As a distinguished foreign teacher at the Board of Education, I had gained special access to many people in town; thus I could hardly expect to excuse myself from various rules because they didn't suit my needs.

"You are a teacher," Mr. C explained as he pulled me aside one day to explain that red means stop in Japan and green means go. "During school or after school you are always a teacher. You must obey the rules."

This was the primary lesson for a new teacher in Japan: the closer you get to the songlines, the stronger the pull of the web.

3

LEARNING THE WAY: THE FIRST DAY IN SCHOOL

Students should be taught to maintain order in their environment, make effective use of time and space, lead regulated lives, and, at the same time, understand the significance of manners and etiquette, and be able to act in any situation.
— Course of Study for Secondary Schools in Japan

A SWEET POTATO SALESMAN reeled past my window at 7:15 in the morning, drawing me out of sleep with his winding refrain about roasted potatoes and his calls to everyone within earshot of his portable grill to have a bright and cheerful day. *Yakii-imo. Oishii-imo.* Soon the pad of students' feet filled the street below and bicycles clanked past on their way to school. At precisely 8:00 the doorbell rang.

"Are you Mr. Bruce?" asked the slightly stooped man at the door as he straightened his tie and eased his hand over his flat-top gray crew cut. "My name is Fuji from Sano Junior High. I am here to take you to school."

Downstairs, Mr. Fuji hurried to hold open the back door of his car like a nervous boy on his first date. I demurred with a slight bow and moved toward the front seat.

"If it's okay with Mr. Bruce, tomorrow you can ride your bicycle, but today I will show you the way. I hope you can fit into the car. It's very small, I know."

The engine screeched as Mr. Fuji tried to push the gearshift into place. He giggled slightly, then gave it one last slap. The car lurched forward with a yelp.

"I guess my car doesn't like foreigners," he said. "Please put on your seat belt."

We followed the chain of children down the main street of Sano and within several minutes could see the high black fence that surrounded the grounds of the school. As we approached, I noticed several dozen students standing in line along the side of the road. They stood quietly in the grass, waited for a car to drive past, and then, as if pulled by a giant rubber band, bowed together at the waist and screamed at the top of their lungs, "*Ohayō gozaimasu!* Good morning!"

"What are those students doing?" I asked Mr. Fuji.

"They're practicing their morning greeting," he said matter-of-factly.

"But they're junior high school students. Haven't they learned how to say good morning?"

"Of course they have. But it's the fall term now, so many of them may have forgotten how over summer vacation. They have to practice again to make their greetings bright and lively. This is the best way to learn."

The school year in Japan, like the business year, begins in April — an old custom that links the opening of the year to the flowering of the cherry blossoms in spring. The first term extends through July, the second term from late August through December, and the third term from January to mid-March. The longest break in the schedule is the three-week vacation in late summer, which many older teachers insist allows their students to forget what they know.

At the door of the school Mr. Fuji dropped his shoes into a small locker and pulled out a pair of white padded bedroom slippers. He gestured for me to follow and offered me a pair of green plastic guest slippers, which pinched my toes and left almost two inches of my heel dragging on the ground. Feeling a bit foolish with green slippers and my first-day-on-the-job blue blazer and red tie, I made a mental note to bring my own indoor shoes tomorrow. At least they were black.

Mr. Fuji ushered me into the teachers' room — the large office where I had performed my formal greeting the previous week — and down a long row of desks to an empty place at the end. "This is Hamano-*sensei*," Mr. Fuji said as we arrived. "Please sit next to him. He will take care of you."

After Mr. Fuji disappeared, the young teacher turned toward me, stuck out his hand, and said, "Hello, I am a freshman English teacher. My name is Kenzo Hamano, but you can call me Denver."

"Denver?" I asked as we shook hands.

"Yes, Denver. I once met an American in my university days. He said I looked like John Denver, so that became my nickname."

"Does everybody call you Denver?"

"Oh no, John Denver is not very famous in Japan. But that doesn't matter. You can call me Denver anyway."

Denver had the same rounded glasses and straight sweep of the hair as his famous namesake. He was slightly short by Japanese standards, but dressed smartly in gray flannel pants, a black blazer, and a pale green Polo tie. His slippers were blue with a slightly raised heel to give him a little more stature.

"By the way," he said, "do you like to sing?"

"Yes, why?"

"Because my students like music and they want to hear your voice."

Around the room other teachers scurried to prepare for the opening bell at 8:30 A.M. A total of thirty-eight teachers worked in this one room, and as in my office at the Board of Education, the desks were grouped together by assignment — all the homeroom teachers for one grade together, the four administrators perpendicular to the others in front. The teachers, ranging in age from twenty-five to sixty, chatted excitedly as they rummaged through their desks and skidded around the room in their slippers, most of them followed by small bands of students tugging at their coattails and asking questions rapidly. "*Sensei, sensei,* what was the homework for today?" "What books do we need for class?" Every few minutes the principal emerged from his office to make an announcement, the photocopy machine started sneezing, or the telephone rang, sending another teacher scuttering across the tile floor.

"Excuse me," said a woman, tapping me on the shoulder. "I hear you like green tea. I made some for you to enjoy."

Mrs. Negishi, as she introduced herself, was a pudgy woman in her late thirties with round white cheeks and black curly hair. As she talked, she bounced up and down on her pink terry-cloth slippers.

"I went to California last year," she said happily. "I visited Disneyland and Hollywood. I brought back pictures for my students. They are very anxious to learn about America from you. By the way, have you been to London? They want to know about that, too."

The opening bell sounded, and the teachers hurried out the door until the only ones left in the big room were Mrs. Negishi, Mr. Fuji, Denver, and I.

"We are going to have a meeting," Mrs. Negishi said. "We hope that you will join us."

Mr. Fuji, the head of the English department, conducted the opening meeting in Japanese. He explained that in my first week I would visit eighteen classes, host the English conversation club, and make model speaking tapes. When I was not teaching, I would make plans for my upcoming classes. Each of the three teachers was responsible for the English classes for one grade. Denver taught the seventh grade, Mr. Fuji the eighth, and Mrs. Negishi the ninth. In addition, each was responsible for a homeroom class.

"My class won the volleyball tournament last summer," Mrs. Negishi interjected. "They are the best in the school."

"As you know," Mr. Fuji continued, "our goal is to teach 'Living English.' " This term was invented by the Japanese government to describe the communication skills it wanted all students to learn. But after announcing this policy with great fanfare, the government realized that most of its teachers did not have these skills. As a result, the Ministry of Education decided to invite native speakers of English into the schools to add new "life" to the language. Under this plan, the foreign teachers would work alongside Japanese teachers in an arrangement the government termed "Team Teaching." In our meeting, everyone agreed that Living English was important and that Team Teaching was the way to achieve it. But once we left the office and stepped in front of the students, we learned quickly what the government did not know: teaching is an awkward team sport.

"Stand up," a student barked as Mr. Fuji and I entered the eighth-grade classroom.

"Ready . . . set," the boy cried again, drawing his classmates to attention.

"Bow."

"*Onegaishimasu*," the class uttered together. "Please teach us today."

The students stood facing the front of the class in eight rows across, six persons deep, like pilings in a pier. Boys and girls were in alternate rows, with the boys in black jackets and straight black pants and the girls in blue blazers and pleated blue skirts. Each student wore a pair of white slip-on tennis shoes and stood behind a simple wooden desk on the parquet floor. The walls of the room were bare except for pedagogic signs taped around the room: "LET'S PUT FORTH GREAT EFFORT," "LET'S EAT ALL OUR LUNCH," and "WHEN THE TEACHER CALLS YOUR NAME, IMMEDIATELY ANSWER, 'HAI.' " Alongside the clock at the front of the room, the carefully scripted school motto reminded the class of the type of students they should try to become:

1. Students who are healthy.
2. Students who study by themselves.
3. Students who are thoughtful.
4. Students who work hard and succeed.

Just beneath this sign, in front of the blackboard, a gray metal desk stood facing the room with the intimidating authority of a judge's bench.

"Good morning, boys and girls," Mr. Fuji said in English when the class had finished its greeting in Japanese.

"Good morning, teacher," they droned, with all the enthusiasm of a deflating balloon.

"How are you today?"

"I'm fine, thank you, how are you?"

He motioned for the students to be seated and turned toward me, obviously pleased that his students had demonstrated such proficiency. "Please give your self-introduction," he whispered in my ear. This was an English class, but Mr. Fuji seemed afraid to use English in front of his students. I took a breath and began.

"Good morning, everybody," I called out, bounding toward the blackboard and grabbing a piece of chalk.

"Good morning, teacher," came the tepid response.

"No, no, no." I waved my hands. "This is not a test. Please do not repeat everything I say."

"Sit up," Mr. Fuji barked in Japanese. "And be quiet."

I cringed at his harsh words but paused in deference before beginning again.

"Good morning, everybody," I said, this time slower. "My name is Bruce. Not *Bu-ru-su*, but Bruce."

Several of the students giggled at my mock Japanese accent. Mr. Fuji moved to the back of the class and rested his arm on top of the lockers.

"I come from Georgia, in the United States of America." I drew a quick sketch of the United States and put a star in the southeast corner. "In Japan, you have *Georgia* Coffee."

Georgia Coffee, a product of Coca-Cola, is a popular brand of canned coffee sold both warm and cold from vending machines all across Japan. The design on the label shows a woman wearing a hoop skirt standing before a large antebellum mansion. As I had learned from the principal on my first visit to the school, Southern gentility still sells in Japan.

I drew a picture of a tin can on the board, and slowly the students caught on. "*Co-he, co-he,*" they screamed. Mr. Fuji stepped forward to silence the swell, but I waved him off and approached the class instead. The students started grumbling as I eased up to a beaming boy in the front with a closely shaved head.

"*Gambate,*" the class shouted, encouraging him to do his best.

"Please stand up," I said to the boy. But before my words had time to bounce off his face, Mr. Fuji had converted my request into Japanese.

The boy rose to his feet.

"What is your name?" I asked.

"Eh?" he said, squinching his face.

"What is your name?" I repeated. "*My* name is Bruce. What is *your* name?"

The boy blanched and recoiled in fear. Silence flooded the room, and he sank back into his seat. Time to try a new student.

"What is *your* name?" I asked a boy sitting in the second row.

"My name is Takuhiro Kobayashi," he said.

"Good. My name is Bruce. Nice to meet you." I stuck out

my arm to shake his hand. The boy looked at it for a moment, then grabbed it and bowed his head. The class erupted into laughter. I asked another question. "Do you like Georgia Coffee?"

The boy turned around as soon as I finished the question and consulted with several friends. Together they began decoding my question. "Coffee, coffee, *co-he*," they mumbled excitedly. "Georgia *co-he*. *Suki desu ka*? Do you like? *Ocha*. No, not tea, coffee." *Thwap!* One student bopped another on the head. "*Hayaku*. Hurry upu! Curazy boy."

After about thirty seconds Mr. Fuji marched over to the desk and translated the question into Japanese.

"Do you like Georgia Coffee?" I repeated in English.

Satisfied that he had understood, young Kobayashi turned back toward me and with all the composure of a seasoned public speaker announced in a loud, clear voice, "Yes."

The class applauded.

"Good. Now everybody repeat, 'I like Georgia Coffee.' "

No sooner had I completed this exchange than Mr. Fuji tapped me on the shoulder. "Mr. Bruce," he whispered again. "This is the end of your self-introduction. We have no more time for this." He promptly returned to the teacher's podium in front of the blackboard and resumed his lesson plan. As I hovered in the back of the room, he reviewed the homework from the previous lesson, in which students had to diagram a series of complex sentences, and then asked several members of the class to recite the week's chapter from the textbook by heart. Mr. Fuji remained at the podium for the duration of the fifty-minute class, following the lesson plan he seemed to have established years ago. He taught the grammar lesson, in Japanese. He had the students write notes, in Japanese. And through this he tried to teach the model sentence, "In the future, I want to become a . . ." No student asked a question of the teacher; the teacher asked no questions of the students. When I interrupted and tried to break down this pattern, the response was always the same: students consulted one another in groups, translated my question into Japanese, then crafted as short an answer as possible.

"What sports do you like?" I asked.

"Baseball."

"What is the weather today?"

"Fine."

"What do you want to become in the future?"

"Salaryman."

These students had learned the rules of English, but they had not learned how to apply them. They could recite an entire chapter by heart but could not tell me the time of day. By the end of the first fifty minutes I was sagging under the weight of my assignment. I was relieved when the bell sounded and the students rose to say good-bye.

"*Kiritsu,*" the young male voice called again.

"*Rei,*" he said.

The students bowed in unison. "Thank you for caring for us today."

"My lesson was very disappointing for you, wasn't it," Mr. Fuji said to me as we walked back to the teachers' room.

"I was surprised that we spoke so little English."

"I'm afraid my English is not very good," he said. "I was not originally an English teacher. I was trained to teach Japanese history, but the principal asked me to switch."

"Was it difficult?"

"Not really. My father was a teacher before the Second World War. He taught me that good teachers teach more than their subjects: they train their students in how to behave. I conduct my English class just like a history class. Students can learn English on their own, I think, but they need me to teach them the Japanese heart."

For most of its history Japan has struggled to preserve its

native heart while its mind looked abroad for new ideas. The country's first public school opened its doors in the latter half of the seventh century, and its goal was to teach students — mostly sons of samurai — to appreciate the arts of China. Over the next several hundred years Japanese emperors founded a series of institutions designed to train officials who would have the skills to manage the imperial court and the taste to appreciate Chinese culture. Over time, as the emperor slowly lost power to the rising warrior class, the imperial schools declined as well and gave way to a new type of institution in Buddhist temples. In 1439 a group of Zen monks established the first of these schools in the mountain village of Ashikaga, in a small, one-room wooden hut that still stands today, about a twenty-minute ride from Sano Junior High.

The temple schools, like the secular ones before them, taught students knowledge from China, in this case the principles of Confucian thought, which Japan had recently embraced. Even after the ruling Tokugawa shogun ushered Japan into a 260-year period of isolation beginning in 1600, Buddhist priests continued to teach the two primary objectives of Confucian thought: administrative skills and cultural appreciation. In these schools, formal ceremonies were stressed, discipline was staunch, and students were asked to memorize texts for comprehensive examinations. As the population flooded to the growing cities, schools in the eighteenth century began to serve not only warriors but merchants and other laymen as well. By 1868, when the shogunate finally lost power to a group of warriors who restored the emperor Meiji as sovereign, Japan was poised to create a new nationwide school network.

With this revolution came a shift in focus from China to the West. In 1872 the new national government announced a plan to form a modern school system and sent emissaries to Europe to bring home ideas. The Japanese modeled their system after the French prototype, with six years of primary school, ten years of secondary school, and four years of university. To encourage studying, the emperor publicly hailed education as the key to developing the "national consciousness" that would help Japan become a rich and powerful nation. Those institutions that had previously aimed at developing government officials began to funnel talented students into industrial fields. Although the government could not at first build enough schools to meet its goals, by 1900 the country had enough facilities to require children to attend school for four years; by 1904 this was increased to six.

As Japan gained industrial strength, its love affair with the West from the 1870s and early 1880s was overwhelmed by a conservative countermovement, which culminated in the Imperial Rescript on Education of 1890. This time the nativist side of the Japanese heart took over from the internationally minded side. By the start of the twentieth century, teachers were promoting a national pride in Japan's military past and a faith in the emperor as a god. Students learned military skills in school along with their reading, writing, and arithmetic. By the late 1930s, Japan, which had so recently admired the cultures of China and the West, was racing toward war against both.

At the close of the Second World War, occupying American officials decided that the extensive network of schools had been exploited by warmongering leaders to indoctrinate children with imperialist ideas. The Allies' solution was to reorganize the schools along the model of the American system, with six years of elementary school, three years of junior high school, three years of high school, and four years of university. Under this plan, children were guaranteed schooling for nine years instead of six, and teachers were told to replace their nationalistic sermons with lessons in "democracy" and "individual freedom." Over forty years later the American superstructure remains in place, but it has proven to be far less American than the Americans had hoped. As it had done repeatedly in the past, Japan adopted a foreign prototype and transformed its alien character by implanting a Japanese heart.

"Good morning, boys and girls," Denver said in English to the class of seventh-grade students after they had bowed and greeted us in Japanese.

"Good morning, teacher," they shouted back in unison like a glee club. Shocked by their enthusiasm, I covered my ears in mock surprise, and the whole class burst into laughter.

"This class is very perky," Denver had said to me on the way to the classroom. "They like English very much. Today I want them to meet an American teacher, so we will have no plan."

I introduced myself as I had in the earlier class, waiting until the part about Georgia Coffee to approach the class. As I walked through the rows of desks, the students squirmed in delight and let out a long, unearthly *oooooh*, like a horror movie come to life.

"Hello," I said to a student in the front of the class.

"*Hai!*" he shouted, popping up from his chair like a piece of toast.

"What is your name?"

The boy did not even have time to turn around before several friends came darting up to the front to whisper in his ear. Within seconds, half the class was shouting his name out loud.

"Matsumoto, Matsumoto," they cried, following the familiar Japanese custom of referring to people by their surnames. "He is Matsumoto."

"Hello, Matsumoto. My name is Bruce. Nice to meet you." I held out my hand to greet him, but the boy looked frightened and dropped his eyes to the floor. Matsumoto, I realized, had never shaken hands with anyone before. I walked back to the head of the class, where Denver was waiting quietly.

"Excuse me, what is your name?" I asked Denver.

"My name is Hamano," he said.

"My name is Bruce. Nice to meet you." I stuck out my hand, and he grabbed it, shaking vigorously.

"In America, we like to shake hands," I said to the class. "In Japan, we bow. Bow, bow, bow." I darted around the room, bending assiduously at the waist. "Hello, bow. Good-bye, bow. Nice to meet you, bow. But in America, we talk with our hands. Hello, shake, shake. Good-bye, shake, shake. Nice to meet you, shake, shake."

I moved around the room, practicing with various students, but something felt incomplete. The students were grabbing my hand now, but still they bowed their heads.

"Everybody, please stand up," I called. "When I was a junior high school student, my father taught me how to shake hands. He said two things were important. Number one: firm grip." I squeezed hard on Denver's hand and made a wide grimace on my face. "Number two: eye to eye." I pointed first at my eyes, then at Denver's. Everybody seemed to understand. But when I moved toward the class, the silence quickly returned. Not only had these students never shaken hands, I realized, but they had also been taught never to look a stranger in the eye. They had learned instead to lower their eyes when greeting a person in order to show respect. As their elder, I could not expect students to look me in the eye on our first meeting. But as their teacher, I had to convince them that showing me respect meant standing face to face and looking me in the eye.

"In America," I continued, trying to resuscitate the situation, "we have several different types of greetings. Sometimes we use a special kind of handshake." I went back toward Matsumoto, lifted one of his arms above his head, and slapped his hand with mine. "We call this a 'High Five.'"

Slowly the students regained their energy and practiced greeting one another with slapped palms and knocked elbows. By now we had used up most of the class with just the basics of greetings.

"If you don't like either of these," I said, moving toward a girl with short-cropped hair who stood gaping at me in the front row, "there is one more way you can try." I stepped up to the girl's desk, lifted her hand in mine, and gave it a quick kiss on the top.

The class let out a dreamy *ooooh* again, and someone called from the back of the class, "Curazy boy. He is a curazy boy."

John Dewey, the American philosopher and educator whose writings held great sway in Japan during the Allied Occupation, once wrote that the most important part of a school is its atmosphere: "The only way in which adults consciously control children is by controlling the environment in which they act, and hence think and feel. We never educate directly, but indirectly by means of the environment." The modern Japanese school is a model of Dewey's vision of control and simplicity. Inside Sano Junior High, students could find few distractions to divert them from study. Rooms were spartan; halls were kept vacant; walls were painted a nondescript beige.

The only decoration throughout the school was the multitude of signs. The walls of the ninth graders, on the third floor, were decked with black and white brochures listing admission requirements for all the high schools in the area, including the time and place of the February entrance exams. The rooms of the eighth graders, on the second floor, were decorated with multiple copies of the class motto, "DIGNITY, HEALTH, AND INTEGRITY." And the walls of the seventh graders, on the ground floor, were draped with calligraphy from a recent writing lesson. In each piece the same two Chinese characters appeared. They read, simply, *Seijin*, "Becoming an adult."

"We must teach our students discipline," Denver explained to me after class, over a lunch of beef stew, cold spinach, and steamed rice. "This is the role of the schools. We are responsible for *shitsuke*."

"What's *shitsuke?*" I asked.

"It means discipline. We have to teach students how to behave properly both in school and out — how to follow rules and develop an honest mind."

"But what about the role of the parents?"

"The principal says they can't be trusted. Some of them may be working, or not care about their children. In Japan, we teachers must be specialists in bringing up children. *Shitsuke* is the heart of our schools."

"But all these signs," I said. "Can they really teach the students?"

"I think they are important," he said. "I can remember walking into my classroom when I was in a bad mood and being reminded that a bright greeting and a healthy attitude are important. I would be lifted from my bad feeling. If my students do not give a bright greeting in the morning, I don't believe they are showing their true feelings. Everyone should be able to give a cheerful 'hello' all the time."

"But what if they aren't in a good mood?"

"They must learn to persevere. If they were one person on an island, they could always say what they think. But they are not. We all must live with other people. This is our Japanese custom."

After lunch I taught my first class of ninth graders with Mrs. Negishi. She began with the same formal introduction as her colleagues but then diverted from the normal rigor into her own personal style. On this day, as on all others, a student was asked to stand before the class and deliver a short speech in English.

"Good afternoon, Mr. Bruth," the young boy said at the start. "Let me say welcome speech for you. In our class there are twenty-one boys and twenty-two girls. Boys are handsome and clever. Girls are beautiful and kind. Today we are very happy to have an English class. We are waiting for you for a long time. We try to catch your speaking, but our English is not very good. We will do our best."

By the time students reach the third year of junior high school, they have been studying English fifty minutes a day, three days a week, forty-five weeks a year, for over two years. Whereas most seventh graders view English as fun and adventurous, most ninth graders see it as a catalogue of rules to be memorized for the high school entrance exam. The public school exam in Tochigi is based entirely on written material and requires no speaking or listening skills at all. Living English may be alive in the hearts of government officials in Tokyo, but it has no life in Sano.

Mrs. Negishi tried to balance this burden with a parcel of exciting exercises. She had trained her students well, so class moved quickly. She began with a vocabulary quiz, in which she flashed a word in English and the students raced to call out the word in Japanese. Next came a word game where she called out the present tense of a verb and one student from each row had to race to the blackboard and write the past tense with chalk. For this lesson on the comparative and the superlative, Mrs. Negishi suggested that we divide the students into two teams and run a drill like a TV quiz show.

"Which is taller, Mount Everest or Mount Fuji?" I asked.

"Mount Everest," a student called.

"Which is bigger, Tokyo or New York?"

"Tokyo."

"Who is faster, Carl Lewis or Mr. Bruce?"

Just as the students were warming to the game, the principal appeared at the door. The students quickly snapped upright in their chairs and crossed their hands in their laps. Sakamato-*sensei* excused himself before Mrs. Negishi and summoned six boys by name. "Please come with me," he announced. The boys buttoned their black jackets with care and followed him silently down the hall.

The game trickled to a close as class ended, and an anxious silence took hold of the school. Downstairs in the main office, teachers stood huddled in small groups, shooing away students who wandered into the room and glancing over their shoulders at the principal's door. Soon Denver appeared at his desk and explained to me what had happened.

During lunchtime, while students were playing in the halls, an eighth-grade boy had called a ninth-grade boy by his name without the honorific *san*. The older student, thinking this impertinent, rounded up some friends and returned to teach the young boy a lesson. The band of six ninth-grade avengers picked up the lone eighth-grade offender, dragged him into a bathroom stall, and began punching him. This continued unabated until a teacher unknowingly walked into the bathroom and discovered the fight in progress.

Back in the office, the teachers waited anxiously for the boys to emerge from their meeting with the principal. Soon Sakamoto-*sensei*, looking gruff and angry, pushed the boys through the teachers' room and led them into the foyer of the school, directly in front of the trophy case.

"Attention," he called out, and the boys jerked into formation.

"You are ninth-grade students," he shouted at them in harsh, bludgeoning speech that echoed through the halls. "You are leaders in this school. It is time to understand your duties and assume your responsibilities." As he talked, the principal marched down the row of boys, slapping each boy on the crown of his head and pushing him down to the ground. As each student landed, he assumed the formal Japanese sitting position by kneeling down on the hard tile floor while resting on his ankles. The students winced in silence.

"You have brought shame on your class and your school. You must be punished. Do you understand?"

"*Hai*," the boys cried loudly, as the signs had instructed them to do.

"Good. Now sit here and think about it."

The principal walked away, giving a last slap to one of the boys, knocking him briefly to the floor, and sending the seventh graders who had been peering around the corner scurrying back to their room. The six students were left on the floor for the next thirty minutes, slowly squirming as their legs cramped in pain. As the other teachers moved in and out of the office, most stopped before the line of students to knock one on the head or whisper some encouragement for all to hear. One particularly ferocious teacher drew a short bamboo pole from his desk and ceremoniously slapped each boy on the back of the neck, like a Zen master rapping young monks to instill determination. The entire faculty body seemed to join in the punishment, and the entire student body seemed to share in the boys' grief. As I watched this public humiliation, I thought of what Denver had said, "*Shitsuke* — discipline — is the heart of our schools."

"Why was there a fight today?" I asked Mr. Fuji as he drove me home from school. On the street, streams of students from school were biking past us, heading home for the day.

"Every year at this time, the ninth graders become very anxious about their future," he said. "Entrance exams are coming up, and they must work harder now. Their hearts are not in balance. Also, the teachers are busy with preparations for the sports festival next month, so they have no time to observe the children. Without the teachers, the children are lost."

Then, without warning, Mr. Fuji started blowing his car horn and pulled suddenly to the side of the road. With a force that belied his age, Mr. Fuji jumped out of the car, crossed the street, and began berating two girls on bicycles. The girls hopped off their bicycle seats and withdrew their white riding helmets from their baskets. They bowed toward the teacher, snapped them into place, and pedaled away again.

Mr. Fuji shook his head and slapped the steering wheel as he sat back in the car. "The students think that once they are away from school, they can do whatever they want. But they know to keep their helmets on at all times. Students today are growing lazy and disobeying all the rules. They are becoming just like Americans. If they are going to be good Japanese, they still have a lot to learn."

4

FACES IN THE DARK:
THE WELCOME PARTY

The apprentice fixed a chair for him amid the banqueters, hung a lyre above his head and led his hands upon the strings. He placed a bread basket on the table and poured a cup of wine. Then each man's hands stretched out upon the banquet.
— The Odyssey, *Book VIII*

AN ENVELOPE ARRIVED on my desk at the Board of Education late one Wednesday afternoon. On the outside was a question: "Have you gotten used to your life in Japan?" On the inside was an invitation to a party. The party was scheduled for that evening; the place was the Lucky Eel Inn; the guest of honor, myself.

After several weeks on the job, I would be officially welcomed into the group of my fellow teachers on that night through the ceremonial ritual of an *enkai*. My office held such social evenings once a month; others held them as often as once a week. These festive functions are as important to office relations as the morning bow or afternoon snack, but one crucial difference remains: at an *enkai* the lubricant is not tea but liquor.

We stopped at the door of the inn to leave our shoes, turning them backward so the toes would face outward when the time came to leave. Inside, a private salon awaited. Tender straw mats edged in violet silk covered the floor; blond wooden *shōji* screens lined with brittle rice paper softened the afternoon sun; a burnt-orange blossom peeked out from a niche in a corner beside the window. The room had the elegant air of a private courtyard in a seventeenth-century romantic novel. The sliding doors and tinted screens, born of that era, were fashioned to shield out formality and provide a den where true feelings could reign — without fear and especially without shame.

The dozen guests sat on large scarlet pillows set evenly across the tatami floor, with two rows of low tables in front of them. Like its cousin the *aisatsu*, or greeting, the *enkai* follows an established form, with a beginning, a middle, and an end. At the start, formal speeches are in order. On this night, Director Kato, a large, dignified man who speaks imperial Japanese but not a word of English, delivered a prepared speech in my native tongue.

"Welcomu, Mista Burusu," he began slowly. "I am bery glad to see you . . ." He hesitated, and the others applauded. "We are bery happy that you have come to live in our city. I hope you have good time here and return to your country in a year and successfully fulfill your duties. Thank you very much."

I was slightly surprised that the director chose my welcome party as the time to talk about my departure, but I was even more surprised by his choice of words, urging me to fulfill my duties. My first thought was to ask, "Duty to whom?" but I thought better of it, realizing that a Japanese person in my place would not ask such a question. Duty comes more naturally in Japan.

After several more speeches, much applause, and the presentation of a bouquet of flowers, Mr. Deputy Director delivered the opening toast. Beer bottles were opened, glasses raised, and all joined in a ringing "*Kanpai!*" — Cheers! The ensemble clinking of glasses signaled the end of formality. What followed was a rare treat for me: a glimpse beyond the public faces of these teachers — the stoic, formal patina that glosses office life — to their uninhibited, private faces.

After the toast we unfolded our legs, cracked open our wooden chopsticks, and stretched our hands upon the feast. The first course consisted of fine slivers of uncooked carp nestled in a bed of shredded mint leaves and garnished with a bright yellow daisy. Other dishes included braised squid with lemon and pepper, poached quail eggs, and batter-fried eggplant with ginger root. Fifteen minutes later the main course arrived: sautéed eel on rice. Eel, *unagi*, has been a delicacy in Japan since the fourteenth century and is believed to have mystical powers ranging from the regenerative to the aphrodisiacal. My officemates were convinced, however, that the thin strips of flaky, filleted meat would be too foreign for my fancy. "This is *very* Japanese," they said.

I bit down on the mushy meat and found, to my relief, that my tongue could take it. I smiled through the slightly pungent taste and assured my colleagues that I too loved eel. They raised their glasses and howled, *"Banzai!"* This cheer, which literally means "May you live ten thousand years," was once reserved only for the emperor, but these days it has a more secular meaning and can even be used on foreigners.

The food disappeared in less than twenty minutes, and the guests turned their attention to the wine and beer. Formal drinking in Japan, like eating, follows a strict code of manners. First, no one is allowed to pour a drink for himself or herself. Seated in the place of honor, I witnessed a virtual stampede as my colleagues tripped over one another to be the first to fill my glass. No sooner had I finished one beer than a new reveler appeared with bottle in hand. After several shots in rapid succession, I realized that sitting in the place of honor all night would quickly leave me incapacitated. I rose, took the offensive, and armed myself with a one-liter bottle of cold Sapporo beer and a small decanter of warm rice wine. My colleagues were equally honored to drink from the foreigner's stock.

After several hours the room reminded me less of a Japanese tea garden and more of a classic bacchanalian den. Crumpled pillows littered the floor, covered by men in assorted reclining poses and women, as always, on bended knees. Empty beer bottles toppled over each other onto growing piles on the floor. The laughter grew louder, the jokes got bawdier, and the tension receded, as if a boulder had been dislodged from a stifled geyser. This was the time when true feelings were released. This was the time, I sensed, to beware. Presently Mr. C led me to the corner and began to talk of my duties at school.

"You must work hard this year," he enjoined. "Your job is very difficult. You must visit many schools and teach many students. I am very sorry about that.

"Also," he said with a nervous laugh, "I don't speak English very well. I only speak Japanese English. The students are the same way. This is very strange for you, I'm sure."

Some of the other teachers gathered around us, slapping my back and filling my glass while he talked.

"But," Mr. C continued, "you must not laugh at your students when they make mistakes. Japanese people have very little confidence, especially in front of foreigners. They are very shy. Japan is an island nation, and we are afraid of outsiders. Arai-*san*, for example, had very little confidence before you came." He gestured at her across the room; she smiled anxiously.

"She had never met a person like you," Mr. C added. "But now she knows you. She knows that you drink tea and eat crackers, just like the rest of us. You must give your students confidence, as you have given her. It will be a challenge for you."

As he concluded his talk Mr. C leaned closer, almost tumbling into my lap. The room grew quiet. For a moment he said nothing, and his face, flushed beet-red from the beer, hung in the air just inches from my own. Then suddenly he leaned back and sighed. "That's the way it is," he said. "It can't be helped."

He lit a cigarette and blew a stream of smoke into the air.

When he turned away I excused myself from the party and retired to the men's room. Poised over a miniature urinal, I had a moment to reflect. What was I to make of this scene? Here was a middle-aged science teacher who had risen to some level of prominence in his career. He had a wife, two kids, two cars, and a large house. He was my boss. Yet he chose this occasion to tell me that he felt insecure around me because he could not speak English and because "Japanese people have very little confidence, especially in front of foreigners."

I believe there is a large measure of truth in his claim that Japanese people often feel shy and guarded around non-Japanese. This is especially true in the countryside. In Sano, people often pointed at me when they saw me on the street. "*Gaijin, gaijin,*" they shouted. "There's a foreigner. He's so tall. He looks funny." In school, students giggled when I spoke in class or went silent when I asked them a question. In a sense Mr. C was apologizing for this behavior. "Japan is an island nation," he had said. "We are afraid of outsiders. It can't be helped." Ironically, at a time when many outsiders view the Japanese as powerful and menacing, many Japanese view themselves as small in size and weak in the face of foreigners. Most respond to this fear of outsiders by escalating their faith in themselves, and especially in their culture.

Mr. C had said it clearly: "We Japanese may not speak English very well, but we have the best wine and the best eel in the world." This message was too explicit to mention over open desktops at work, so my boss had waited until after hours and behind closed doors at our evening *enkai*. Though we had left the office long ago, we were still conducting official business. An *enkai*, I realized, was much more than a public spree; it was a time for sharing private thoughts that prudence deterred one from saying at work.

When I returned to the room, the party had advanced to the singing stage. A black stereo case with a portable microphone had materialized in one corner of the room. After a little perfunctory cajoling, each of the men paraded to the front of the room to regale his colleagues with dripping love songs and paeans to the beauty of Japanese women. "It's autumn; the leaves are turning; my heart falls for you." The crowd responded with appropriate hoots and whistles as each of the men — though none of the women — took a turn at the magic box that could turn even the most grating voice into a deep, throaty swoon. Each of the men was required to perform this ritual, no matter how grating his voice truly was. Though the acts appeared nonchalant, I have known Japanese men who practiced for weeks before an important *enkai* so that they would not embarrass themselves before their colleagues.

The anticipated moment arrived as the newcomer rose to perform his last and perhaps most important initiatory rite. I took the microphone, and the crowd began to chant, "Sain-su. Sain-su! We wantu sain-su." After a moment to decode this Japanese English, I responded to the call.

"Oh when the Saints (bum, bum, bum, bum) go marching in, Oh when the Saints go marching in, Oh how I'd love to be in that number, When the Saints go marching in . . ."

The crowd loved my mock Louis Armstrong and begged for more. Frantically stretching the banks of my limited repertoire, I hit on a sure-fire winner, a participatory song that would satisfy their appetite for rowdiness. My next number, I explained, was a fight song used by my college football team. "Bulldog. Bulldog. Bow! wow! wow! E-li Yale. Bulldog, Bulldog, bow! wow! wow! Our team will never fail . . ." Enchanted by the lyrics, my boss, my section chief, and even the director of the Ansoku Education Office began chanting and barking so loudly that John Harvard must have turned over in his grave, ten thousand miles away in Massachusetts.

So there I stood, towering 187 centimeters above the floor in my stocking feet, leading my harem of 170-centimeter drunken schoolteachers in a howling ensemble of "Bow! wow! wow!" and discovering for the first time what my contract meant by urging me to promote mutual understanding and cooperation.

More important, I had achieved with this song a sense of parity with these teachers. Not only could I use chopsticks and bow, but I could also make a fool of myself in front of everyone. Now the ceremony could end.

The final act of the *enkai* seemed to emphasize this group feeling. Just as the evening opened with an ensemble toast, it concluded with an ensemble chant. Everyone joined in a large circle, bent down, and then rose slowly with a growing cheer, "HeeeeeEEEEEY!" *Clap! Clap! Clap!* The formal party concluded on this note, and we all filed toward the sliding paper door, where our shoes lay patiently in wait.

If the party had ended there, I would have returned home laden with flowers and aglow with a queasy yet warm feeling. As it turned out, my boss had other plans. He intended to serenade me long into the night and take me deeper into the "floating world" of Japanese after-hours clubs. As the guests piled into cabs in the parking lot, Mr. C whisked me into a separate taxi. (The most impressive feat of the evening proved to be the extensive precautions exercised to avoid drinking and driving.) Having ditched his boss and the office ladies, Mr. Cherry Blossom pulled closer to me on the linen cover of the backseat, put his arm around my neck, and declared, "Now, we drink."

The car stopped in the center of town, and we proceeded by foot to the mouth of a narrow lane teeming with flashing lights and neon: "GIRLS! GIRLS! GIRLS!"; "PLAYBOY"; "SEE A SHOW." Even the porn was in English here. We dodged a cluster of dubious women who surged forward calling my boss — "*Sensei, sensei*" — and walked past the Playboy club to enter a quaint little *kara-oke* bar, bathed in a fog of artificial candlelight and unfiltered smoke. *Kara-oke*, a hybrid term meaning "empty orchestra," is to the Japanese what epic poetry was to the Greeks. The name refers to the prerecorded music without words that provides background for bacchanals to sing themselves silly in bars and backyards all across Japan. *Kara-oke* pubs, which are more numerous in Japan than everything but rice paddies, combine the love of singing with the need for nightclub recreation.

Although the bar was tiny, about the size of a large Toyota, it boasted a giant-screen television suspended from the ceiling, a five-component stereo system, and an exhaustive video and tape library containing over eight hundred songs. The wall behind the bar was lined with dozens of half-empty liquor bottles known as "Bottle Keep" whiskey. For around $100, the faithful customer could purchase a twenty-six-ounce bottle of American whiskey, which would bear his name and be used to fill his guests' glasses when he came to visit.

"Mama," Mr. C called to the hostess behind the bar, "pour this man a drink." He mounted a stool at the bar as the *kara-oke mama* turned on her charm. She was older than Mr. C and dressed in coquettish style, with her hair pulled tight in a bun, thick make-up on her cheeks, and lime-green eye shadow catlike around her eyes.

"I am so honored to receive you," she said with a smile. The *mama* took a special interest in the newcomer. "Oh, the honorable foreigner speaks Japanese so weeelll," she said with a subtle flutter of her eyelashes. She followed this with a steady stream of food to test my dexterity: here some sliced ginger, there some fermented soybeans, finally some radish pickles. "Oh, the honorable foreigner uses chopsticks so weeelll." Unlike my admirers at the *enkai*, she took my hand with this last comment to caress my honorable foreign fingers, which she said must be so tired from using chopsticks all day. As she served the food and fondled our egos, the hostess paused intermittently to refill our glasses with whiskey. She performed the entire routine with a grace and flourish rivaling that of any three-star-restaurant maitre d' in Paris. This *mama* was clearly a seasoned professional.

The eating and drinking soon eased into singing, and the

kara-oke came alive. The hostess got out three large pink note-books that contained the title, the opening line, and the musical key of every song in stock. A man at the end of the table, who had been silent until then, suddenly perked up when the micro-phone was offered. Ever watchful, our hostess had chosen him first. As his voice echoed in the artificial acoustics of the stereo, Mr. C came alive as well. "Hey, Mama," he called, "let's dance."

The two met at the center of the floor, and he took her by the waist for a slow spin, resting his head on her shoulder and pulling her close. Everyone eyed the dancers intently as Mr. C squeezed her on the rear and ran his hand up under her dress. When the song was over he returned quietly to his seat. He winked at me through his whiskey glass.

After several more songs, the microphone was passed to me and the video screen was kindled. Welcome to *kara-video*, the Japanese nightclub version of Follow the Bouncing Ball. The *mama* selected "Love Me Tender," and before I had time to protest, the lyrics appeared on the screen, superimposed on a picturesque scene of two lovers, very blond and very American, strolling on a sunset-drenched beach. Making my way into my best Elvis bar-itone, I was increasingly distracted by these two lovers as they moved from the beach into the bedroom. Suddenly they stripped off their clothes, fell naked on the floor, and began making love — right there in the middle of the bar. As their bodies writhed with pleasure and pain, the men at the counter started cheering, and I was left standing bewildered, trying desperately to concentrate on the lyrics that kept crawling casually by.

At the end of the song, Mr. C suggested that I take a spin with Mama, but I apologized and insisted I must be getting home. Disappointed but dutiful, he followed. The bill for two people, for under two hours, came to $125.

I said good-bye to Mr. C at the door of a taxi; his voice, still droning "Love Me Tender," echoed through the empty streets. As I walked home with my bouquet of roses, past the barbershop and the vegetable stand that I passed every day, the storefronts appeared different in the shadowy light, as if somehow unreal. I wondered if these buildings, like the men in my office, had faces that altered from day to night.

For several weeks after the *enkai* I was haunted by the people I had met that night, the unrestrained men lurking behind my courteous colleagues at work. I longed to know which were their true faces and how they balanced these dual lives. When they bowed at me respectfully in the morning, did they still hear me shouting "Bow! wow! wow!"? When they humbly addressed a woman in the office, did they visualize her making love on the beach?

I sought answers from friends. Denver, an *enkai* enthusiast himself, suggested that these two faces were necessary devices to preserve a common group. Private behavior, he said, remains in the realm of the private and rarely reflects on one's public persona. There is no shame in making a pass at a *kara-oke mama* or insulting a co-worker at an *enkai*, he said, because no one "remembers" at work the next day. In many cases the company or school will even foot the bill for this kind of group release to guarantee peak performance during the day.

Morality, as I was learning, is set primarily by society as a whole and not by individuals in Japan. If a group condones a behavior, such as drinking or having an affair, then individuals are allowed to indulge. In this world shame is more powerful than guilt, because people's actions are tempered less by fear of internal torment than by the threat of group disapproval. I marveled at the way in which my colleagues could easily separate public conduct, known as *tatemae*, from private desires, known as *honne*. Co-workers who were rude to one another in the bar would be civil the next day at work; men who had been open and relaxed in the bath would be formal and rigid when behind a desk. Although I tried, I was unable to make this separation. When I saw my boss every day, I remembered.

PINK HATS AND HARDENED HEARTS: THE SPORTS FESTIVAL

The child's soul should not learn to feel pleasure and pain in ways contrary to the law, but feel pleasure and pain in the same things as the aged. But since the souls of children cannot bear earnestness, efforts to achieve this are spoken of as "play," and practiced as such.

— *Plato*, The Laws

THE FIRST TIME I SAW the students of Sano Junior High School practice their mass dance for the annual October *undō-kai*, or sports festival, I thought I had discovered the secret to Japanese group harmony. There before my eyes were 690 students dressed in matching white uniforms with pink and blue hats, dancing, chanting, and marching in perfect unison, in what seemed like the ultimate expression of group cooperation and togetherness. The students formed elaborate human constellations and shifted in and out of formation quietly, efficiently, and without complaint. As I watched in awe from the roof of the school, the principal next to me glowed with pride. I could already hear him boasting to the PTA, "Oh, the foreigner was so impressed."

Japanese schools have been widely chided in the West in recent years for raising a population that works too hard and plays too little. In one sense this is true: Japanese children attend school an average of 240 days a year, compared with 180 in the United States. But the real difference between Japan and the

West is not that Japanese schools teach their students how to work but that they teach them how to play as well. From teaching calligraphy after school to giving swimming lessons on Sunday, schools fill the daily schedule of Japanese students with a panoply of activities that teach social and cultural skills. Classes that would be considered extracurricular in the United States are thought to be quite curricular in Japan.

Sano Junior High School sponsors at least two major events in each of the three terms of the year — a sports festival and a culture fair in the fall; a basketball tournament and a graduation ceremony in the winter; a volleyball tournament and a school trip in the spring. The *undō-kai*, which coincides with National Sports Day, is the centerpiece of the fall trimester. Every October, over two thousand parents and community leaders gather on the school grounds to witness the spectacle of the mass dance that inaugurates the day. After the show, parents join in an afternoon of relay races and games, such as the egg toss, the tug of war, and something called the eight-legged run, in which four students stand one behind the other, tie their left legs together with one sheet and their right legs with another, and then run together as one. Over the years Sano Junior High has developed an outstanding reputation in the region for its well-manicured campus and its well-choreographed mass dance.

Early in the week, as I watched students practice for the show, entitled "Time Travel," I marveled at the precision and beauty. The students marched around the field in a maze of geometric formations and intersecting braids that would have been worthy of any Big Ten marching band or Super Bowl halftime show. In one number, the students thundered their feet on the dirt and folded their hats inside out in a flurry of pastel shimmers as they moved into a three-dimensional replica of the emblem of Tochigi. The square design included a pointed leaf to stand for the thick chestnut forests, a circle for the famed Lake Chuzenjiko in Nikko National Park, and two diagonal stripes for the prefecture's two most notable achievements: the Tohoku bullet train and the Tohoku Super Expressway. The climax of the show came as the students formed a set of concentric circles in the center of the field to symbolize the cycles of time. The youngest students sat on the ground, linked their arms, and swayed like trees in the wind; older students stood behind waving red hula hoops in a circle; and ten of the star female athletes, dressed in shimmering blue and yellow leotards, ascended a twenty-foot scaffolding tower, leaned backward over the rail, and jiggled fluorescent pompoms. Over the loudspeaker a souped-up version of "Let It Be" blared across the town.

The boys and the girls often performed different actions in the course of the show. In one routine, the students moved into an enormous drill formation to demonstrate the physical skills they had honed over the previous year. The boys stripped off their shirts, snapped into and out of aikido martial arts positions, and jumped from one another's shoulders. Meanwhile, the girls stood meekly behind waving maroon and gold banners. Next, boys from each class built a magnificent six-person-high pyramid. At the sound of a large bass drum, all twenty-one pyramids crashed down at once in an impressive display of synchronization. As the boys tumbled into a massive heap, the girls bent slowly at the waist like wind-up dolls and touched their flags to the ground. The rhythm was striking, but I wondered about the segregation. Turning to the principal beside me, I asked in my best naive, inquisitive tone, "So why do the boys get to build pyramids and the girls only wave banners?"

"Well," said Sakamoto-*sensei*, the man who had earlier suggested that wives in Japan were like slaves in Georgia, "the boys must learn to be powerful and strong, and the girls must be pretty and graceful."

"I see." I hesitated but decided to plow ahead. "But can't girls be strong too, and can't boys be graceful?"

"Oh no," the principal said with obvious shock. "Girls must be girllike, and boys must be boylike. That's the only way."

After observing the dance rehearsal from afar, I moved

down onto the field to talk with the students directly. As they milled around the dirt field, waiting for the music teacher to fix the tape or the choreographer to make a correction, I asked a group of ninth-grade girls how they felt about the pyramids.

"We hate this part of the show," they said without hesitation. "All we do is dance, while the boys get to show off."

"The pyramids look nice," one girl said, "but we want to build them, just like the boys."

"Can you?" I asked.

"Yes," she insisted. "We have to make them in gym class, but we aren't allowed to make them when the parents come to visit."

Startled by their frankness, I moved to a group of boys to ask how they felt about the mass dance.

"It's very beautiful," they dutifully answered.

"Yes," I said, "but do you enjoy it?"

After checking over their shoulders to ensure that no teachers were in earshot, the boys drew closer together.

"We don't really like it very much," one confessed. "We have to practice every day for four months."

"When the pyramids fall down, we get hurt," said another. "I broke my arm during last year's festival."

I asked both the boys and the girls what I assumed to be the next logical question, "Did you tell a teacher how you feel?" and I received the same definitive answer: silence.

Plato once wrote that boys and girls should dance together while they are young so they can understand their gender roles before they become adults. In a curious way, the modern Japanese *undō-kai* serves the same purpose. It offers up traditional values of "boylike" boys and "girllike" girls, as if the mere act of planting these ideals on the playing field would guarantee that they take root in the students. But the children on that playing field had different ideas.

Paradoxically, women were once venerated in Japan: legend holds that the Japanese imperial line was descended from a woman, the Sun goddess Amaterasu Omikami. But feudal lords over a thousand years ago overturned the legacy of respect and developed an exacting set of laws that subjugated women to men. Confucian teachings adopted later dictated that a woman should obey her father in youth, her husband in maturity, and her son in old age. While women were restricted to the home, men were allowed more freedom — both social and sexual. In time a dual image of women took hold in society as men sought stability and status with the woman they married while acting out their sexual fantasies with pay-as-you-go women tucked away in illicit neighborhoods, known colloquially as the "water trade." Men looked at women as just another part of their segregated public and private selves.

This schism remains in force today. Even though occupying Americans included an equal rights amendment in Japan's postwar constitution, equality never took hold. The Japanese school system for the last fifty years has essentially tried to produce girls who will serve as successful wives and mothers for working men. In the classroom, boys' names are always called first when attendance is taken; outside of class, events like the sports festival ensure that boys and girls follow different paths. The results of this gender training are dramatic. In institutions of higher education, men outnumber women two to one, with the vast majority of the women in less prestigious two-year colleges. Fields of interest vary greatly as well. Ninety-nine percent of engineering majors in universities are men, as are ninety-four percent of law students and almost ninety percent of scientists. Conversely, the most popular subjects for women are home economics and literature.

But times are changing. Polls show that three quarters of women still view a good husband and successful children as the most important goals in their lives, yet these same surveys also show that more women are staying in jobs longer and even continuing to work after marriage. Among younger girls, like those I met in junior high school, the differences are becoming even less defined. Girls rarely use the more polite, softer speech expected of women, even when speaking to teachers, and they use some previously male-only vulgarities when teasing the boys in their class. One of the problems these girls face is meeting the expectations of older teachers whose opinions were formed a generation ago. But these days, as a mark of the evolution, an equal number of girls and boys are willing to stand up in the middle of English class and announce with some conviction, "In the future, I want to become a doctor."

The day after my conversation with the students, as I was talking with Denver in the teachers' room, the principal stopped by my desk to ask my impression of the show. I told him honestly that I thought the mass dance was beautiful but that many of the students seemed not to enjoy it.

"Well, the students don't like English much either," he snapped. The *undō-kai* is like a test for physical education class, he said, just like a math test or an English test. "Therefore, students are not expected to enjoy it."

"But you don't invite parents to watch a math test," I protested.

He paused, and a smile came over his face. "Mr. Bruce," he said, "I think you are very curious about our Japanese school. You want to help the students very much. So do we. But students cannot always do things that they enjoy. If Japanese students don't like something in school they must learn to *gaman*."

This word, which can roughly be translated as "endure" or "persevere," is one of the primary pillars of education in Japan. "If you want to learn how to be a good Japanese," junior high school students are often told, "you must learn how to suffer." Students are coached to *gaman* through difficult tests, long lectures, even an occasional bad lunch. It seemed only natural that the teachers considered the biggest festival of the year a prime time to teach the value of perseverance. The fact that none of the students actually complained demonstrated how much they had already had their rebellious instincts dulled by the endurance training of *gaman*.

Yet I was still unsatisfied. I explained to Denver, the principal, and a growing horde of teachers that I understood the value of teaching discipline and cooperation, but why couldn't they just build four-person-high pyramids so that the boys didn't have to fall so far? And why couldn't the girls have a chance to show off their skills as well? Sakamoto-*sensei* shook his head in resignation and mumbled a brief reply.

The other teachers nodded in agreement and hurried back to work as if the principal had revealed a state secret.

The annual fall sports festival, he had said, was not the tribute to harmony or hard work that I had originally imagined, but the "primary wing" of the school — in short, an elaborate propaganda display. When measured against the favorable opinions of two thousand PTA members and the chief of police, all of whom adored watching girls wave banners and boys tumble from atop human pyramids, half a dozen broken arms and some disappointed hearts were trivial. The students were forced to *gaman*, I thought, while the aged basked in the glory.

As the principal walked way, he abruptly turned and shouted back at me, "Don't you have such things in America?"

"Sure, we have festivals," I said, "but not ones that the students don't enjoy."

He thought for a moment and replied, "Well, that's the difference between Japan and America."

Perhaps he was right. Japanese teachers seemed more willing to overlook the possible discomfort of their students if the school felt an activity worthwhile. The students, in turn, learned early to squelch their true feelings.

Still, I had learned that the students did not always submit mindlessly to such reduction. Away from the teachers, alone among themselves, this generation was already nourishing the seeds of discontent that might flourish later on. Though they had chosen at this time to keep their mouths shut, I could hear them whispering that this would not always be true. Underneath the veil of harmony lay a bunch of scraped knees and a host of budding souls.

6

GOING INTO THE COUNTRY: FALL IN THE CHESTNUT BASIN

In the blasts of wind
chestnut burrs
skitter over the ground,
burr-headed village boys
hot in pursuit.

— *Kanko Akera, "Autumn Wind," c. 1780*

TOCHIGI PREFECTURE is famous for wind.

When I first arrived in Sano, in late summer, the mercury hovered around 40 degrees Celsius (over 100 degrees Fahrenheit) and almost everyone greeted me by saying, "Welcome to Tochigi. We have strong wind." At first I thought this a joke. After all, was the most interesting thing about Tochigi its wind? But as the mercury began dipping toward zero and the leaves started tumbling from the trees, I learned quickly that wind is not a subject for joking in Japan — especially not in autumn.

Tochigi, which is named for the horse chestnut trees that blanket its hills, is the Iowa of Japan. It sits in the geographic center of the country, near the crease of mountains that divides the main island of Honshu, and halfway between the northern island of Hokkaido and the southern island of Kyushu. According to the governor, Tochigi boasts the average population, the average income, and the average "easygoing" lifestyle of Japan. In a country where blending in is at a premium and where ninety percent of the people call themselves middle class, being the middle of the middle, the most undistinguished, seemed to be something worth bragging about.

"Of course Tochigi is quite backward," a friend of mine from Tokyo told me when he learned of my new home. Indeed, to most people who live in urban Japan, any place outside of Tokyo, Osaka, or several other large cities is considered to be *inaka*, roughly "way out" and "way behind." Tochigi, which is fifty miles and fifty years from Tokyo, is far enough away from the capital to be considered *inaka*, but close enough to be part of the region surrounding Tokyo known as the Kanto Plain.

Japan's original capital was located in the Kansai, or Western Region, but in 1603 the Tokugawa shogun moved his royal court to the Kanto, or Eastern Region, to be nearer his military base of support. This shift brought the capital for the first time under the shadow of Mount Fuji, the cultural equivalent of Mount Olympus, Mount Sinai, and the seven hills of Rome all rolled into one. Woodblock prints from less than a century ago show Mount Fuji clearly visible from the capital city of Edo, later renamed Tokyo. In these images the silent, pale face of the mountain looms over the crowded shops and teeming alleys of the pleasure districts, where merchants and peasants sought private diversion in sake shops and geisha salons. As feudal Edo evolved into modern Tokyo and the shallow wooden storefronts gave way first to concrete and stone façades and later to soaring monuments of glass and steel, Fuji was gradually obscured. These days, all one can see in any direction from the heart of Tokyo are silhouettes of high-rise buildings and the silent beacon of neon — the new symbol of Japan — flashing, selling, alluring, warning.

Today, Tokyo's urban bulk fills most of the Kanto Plain, stretching outward from its hub in Tokyo Bay in a dense mesh of highways, railways, industrial haze, and commuter angst. On the outer reaches of this basin, the buildings become shorter and the houses farther apart. Pine trees and chestnut forests line the hills once again. Here, at the gateway to the northern mountains, Tochigi begins.

And so does the wind.

Every year, about the time of the autumnal equinox, a great Siberian bear dons his winter cap, yawns, and exhales an enormous swell of cold, dry air that comes sweeping across the Sea of Japan, through the mountain pocket known as the Snow Country, over the alpine spine of Honshu, and into the valley of Tochigi, igniting the trees of autumn. Tochigi, with its abundant trees, varied terrain, and famous wind, is the essence of autumn in Japan, a time so significant that its beginning is marked by a national holiday. Above all, autumn is a time for excursions. This tradition, which goes back more than a millennium, is described in the eleventh-century romantic novel *The Tale of Genji* and, in the seventeenth century, by Japan's most legendary classical poet, Matsuo Bashō:

> *Bidding farewell,*
> *Bidden good-bye,*
> *I walked into*
> *The autumn of Kiso.*

As with other things Japanese, an excursion into the mountains to view fall colors has a proper form and even a proper name, *momijigari*. In Kanto, and perhaps in all of Japan, the ideal place to seek this communion with nature is Nikko National Park, a cultural landmark and famed retreat perched in the northern mountains of Tochigi. The Tokugawa shoguns thought so much of this place that they built an elaborate mausoleum shrine here to honor the founder of their clan, and made their retainers around the country pay for its upkeep. Nikko — which literally means sunlight — ranks near the top of the wonders of Japan and is a cultural mecca of such importance that thousands of students from hundreds of schools make an annual pilgrimage to the park.

On the second Sunday in October I made this trek with my new friend Cho and his girlfriend, Chieko, and learned a local saying, "Never say *kekko* [I'm satisfied] until you've seen Nikko."

Although my boss had called several days before we were to go to Nikko to warn of the crowds and to urge us to get started well before six A.M., Cho arrived at my apartment close to seven. His girlfriend was already in the car. "I hope you don't mind if I'm late," he said, "but I know a special short cut."

Cho is an altogether autumn person — warm, colorful, and calm. He is also independent. After our first meeting during my opening-night bath, when Cho rescued me from the circle of middle-aged men all trying to develop a meaningful "relationship without clothes on," we had become fast friends. Cho, whose name means chief — as in *kachō*, section chief — taught elementary school in Tochigi and lived with his parents in the musty granite hills of Kuzu, a small mill town just north of Sano. Like Denver, he is short by Japanese standards, with straight charcoal hair that hangs over his eyes, a dark complexion, and a broad, toothy smile. Arai-*san*, the head secretary in my office, once described Cho as having a "warm heart," a high compliment in Japanese used to describe someone who displays an unusual peace of mind and sensitivity. His warm heart blends with a cool insight, which he applies to others as well as to himself. "I was afraid to talk with foreigners before I went to college," he admitted on one of our first meetings, "but I had little confidence then. Now I have more."

University offered Cho, like most Japanese young people, a rare reprieve from academic and societal pressure. Coming after a childhood of rugged preparations for entrance exams and before a lifetime of commitment to a job, college for most is a time of leisure. "We Japanese don't study in college," he explained. "We don't do homework or go to the library. We only play."

After going to high school in Sano, Cho attended a prestigious private university in the heart of Tokyo and "played" with other students in the international exchange club. This group of global savants held drinking parties in chic bars in Tokyo; they jetted to Manila, Cairo, and Paris to meet with other student groups; and they hosted visiting Americans in their tiny dorm rooms at night. As a reminder of these years, Cho came dressed for Nikko in blue jeans, penny loafers, white socks, and a cardinal-red sweatshirt embossed with the emblem of his university club.

With his golden academic pedigree and family background — his father was the principal of a local elementary school — Cho could easily have gained access to the upper echelons of some company or government bureaucracy. He chose instead to become a teacher. At first I wondered why he returned from the fast track of Tokyo to the foothills of Tochigi, but over time I realized that the answer was simpler than I had imagined. As the first son in his family, Cho inherited the responsibilities of filial duty, which all but required him to tend to his parents after his father retired. But Cho had positive reasons for returning home as well. He wanted to bring what he learned from his university and from traveling abroad to the children of the *inaka*. Although his salary from the state was less than half of what most of his college friends earned, Cho liked being with students and encouraging them to dream of the larger world. I once asked him if he found it difficult being an elementary school teacher in the same town as his father. Didn't the old teacher ever take his son aside and say, "Well, son, let me tell you a few things about teaching . . ."?

"He does," Cho said, "but I don't listen. I want to develop my own way."

This desire to find his own way in the face of an established tradition was the central struggle in Cho's life. In returning to Kuzu he had given up much of his personal freedom, but he was not willing to give up his independence of mind as well. Cho and

I often discussed this dilemma. Could he live in conservative Tochigi and still act in a Tokyo way? We deemed his personal quest "Cho-*dō*," in search of the "Way of Cho."

The trip from Sano to Nikko lasted a little over two hours. We followed the Tohoku Super Expressway north for about fifty minutes, then switched to bumpy rural roads to cut time and save money from costly tollways. Unlike the monotonous industrioscape that surrounds present-day Tokyo, Tochigi is softened by rural charms. Along backwoods highways on this mid-October morning, children picked sweet pears while old women and men with rounded shoulders and pointed hats tied sheaves of rice on six-foot poles to dry in the midday sun. Every twenty minutes, a small town would interrupt the farmland, and a cluster of one-room textile mills and food processing plants would appear along the road. This mix of family farms and lightweight industries feeds most of the people in Tochigi but keeps them always one step removed from the prosperous path of the nation.

"Tochigi has more cars per person than any other prefecture in Japan," Cho said as we drove through a particularly bumpy town. "We also have more accidents."

"Why's that?" I asked.

"Because our roads are so bad. Just look at them. If I were governor of Tochigi I would repave all the roads and make them safe for cars."

Just before ten o'clock we finally passed through the last country town and arrived at the entrance to the park. Though we had succeeded in beating the tidal wave of tourists from Tokyo, we did run into a flock of early-bird leaf-seekers from the countryside. The true cause of congestion, however, proved not to be the cars but a bumper-to-bumper battalion of motorbikes. Lured by the famous alphabet highway of Nikko — a treacherous mountain pass with four dozen hairpin turns each named for a letter of the Japanese alphabet — hundreds of bikers had gathered at the base of the mountain for a mass ascent. The riders were all perfectly suited in his-and-her leather outfits — pink and yellow for the ladies, black and red for the gentlemen — with matching helmets and monogrammed saddle bags. Not every member of this crew could ride a motorcycle, and still fewer would manage to make it all the way through the curves from *ah* to *un*, but at least everyone looked the way a biker was expected to look.

Just as the bikers all conformed to a dress code of sorts, I noticed a pattern among some of the more conventional nature-lovers we met. Many of the men sported finely tailored tweed knickers buckled just below the knee, along with an English hunting cap and a carved walking stick. Birdwatchers with telescopes on tripods were decked out in camouflage jumpsuits. I could understand the need to make a distinction between work clothes and play clothes and the emphasis placed on the proper equipment, but a person no more needed a hunting cap and knickers to view the leaves around Nikko than I needed a kimono and samurai sword to visit a feudal castle. The importance of costume, of extending the limited experience of viewing leaves or riding a motorcycle into a complete personal transformation, is part of the larger commitment to form which permeates Japanese culture. Viewing leaves is something to be done not casually but carefully, according to plan. This tradition — like the wind — dies hard.

The legend of the wind in Japan goes back as far as the allegory of the sun. In the beginning, legend says, there was chaos. Through a gradual shifting of particles, heaven and earth were created and various deities came into being in what was known as the Plain of High Heaven. Two of these deities, the brother and sister Izanagi and Izanami, briefly descended a bridge to earth and begot the Central Land of Reed Plains, which later became the islands of Japan. After Izanami died in a fiery accident, Izanagi tried to purify himself and in the process spontaneously gave birth to a daughter through his eyes and a son through his nose. The daughter, Amaterasu, became goddess of the Sun and ruler over heaven. Her brother, Susano, became god of the Wind and sovereign over the sea. But the stormy Susano was jealous of his sister's appointment as leader of all the gods, and in an act of vengeance he wreaked havoc on her territory with storms that ripped through the fields of heaven. The Sun goddess was so distraught that she sought refuge in a cave, thereby eclipsing the world into darkness.

The older deities gathered outside her cave and devised a scheme to lure her out. They placed a mirror and a jewel in the branches of a sacred tree and erected a perch with a cock on top (this simple post-and-lintel perch, known as a torii, later became the principal symbol of the Shinto faith). One of the goddesses performed a ribald dance that the others applauded boisterously. Roused by the noise, Amaterasu peeked her head out of the cave, caught sight of the jewel and her own reflection in the mirror, and emerged from her seclusion. Light once more filled the world.

As punishment, Susano was expelled from heaven and banished to the western shores of Japan; thus the Wind god and his descendants are considered to be the country's first inhabitants. But despite this distinction, they are not given credit for beginning the imperial line. That honor belongs to the Sun. The Sun goddess, after receiving a guarantee from the other gods that the descendants of Susano would submit to her rule, sent her grandson Ninigi to seize control of the Central Land of Reed Plains. In 660 B.C., a great-grandson of Ninigi successfully completed an expedition across the country and was rewarded with the title of emperor of Japan. (The current emperor, Akihito, who ascended the Chrysanthemum throne in 1989, traces his heritage back over twenty-six hundred years in what is claimed to be an unbroken line to this man, the emperor Jimmu, and through him to the goddess of the Sun.)

The jealous uncle, the Wind, has been fuming ever since.

·   ·   ·

After motoring our way up the alphabet road, we broke for lunch at a spot overlooking Kegon Falls, a three-hundred-foot waterfall that cascades over a sheer rock face. Cho's girlfriend, Chieko, began setting out a lunch of chicken strips, rice balls, and broccoli spears which Cho's mother had prepared. Chieko's servility surprised me at first. Like Cho, she was an elementary school teacher. She had hair that brushed her shoulders and short bangs that hung down over her forehead in curls. Her skin was pale, her smile slight, her voice barely a whisper. Nearing twenty-six and the end of her narrow window of eligibility for marriage, Chieko had clearly set her sights on Cho. But the two had known each other only for several months and had been boyfriend and girlfriend for less time than that.

Cho and Chieko, who had met through work, were what the Japanese call a love match, as opposed to an arranged set-up. In the more traditional style of courting, still used by about half of all couples today, a go-between sends glossy photographs and genealogical data to both sides of a proposed match, and then chaperons a meeting between the man, the woman, and both of their families. After this meeting each party decides whether to proceed to a formal courtship, which usually leads to engagement. In love marriages the procedure is considered less strict, although in most cases the courtship is not much changed. In this relationship, Cho would eventually decide whether to propose. Chieko, meanwhile, tried to be pleasant, attractive, and attentive to show that she would make a good wife and companion. Accordingly, she came to our leaf-viewing dressed in a green miniskirt, a white silk blouse, and a blue scarf, and she wore red lipstick and pink blush on her cheeks. This was her uniform, no less fastidious than the bikers', and she looked as if she had just stepped off the pages of a woman's magazine from under the headline "HOW A YOUNG LADY SHOULD LOOK IF SHE WANTS TO CATCH A MAN." She had achieved the look but did not seem to be having a good time. Through the course of lunch, the sights, and the drive to and fro, Chieko said almost nothing.

After lunch we walked around the shores of Nikko's famed Lake Chuzenjiko. At first the path was flooded with bikers, bird-watchers, and small bands of uniformed schoolchildren, but as we continued farther around the lake, the crowds thinned and we were able to behold the atmosphere in a rare moment of privacy. As we wandered in the woods, Cho explained the evolution of the *momijigari*. Certain natural phenomena, because of their splendor and singular beauty, developed almost a religious significance in ancient Japanese culture, where Shinto beliefs held that nature was the home of spirits who lived in the water, the land, and the trees. The mysterious transformation of green leaves into fiery reds and frosty yellows around the time of the harvest every year inspired awe among superstitious farmers. Just as a protocol evolved around making tea (*Sadō*, the Way of Tea) or painting calligraphy (*Shodō*, the Way of the Brush), so a proper form for viewing nature eventually evolved. Modern science has only enhanced this obsession. Every night in autumn on the evening news, the major television networks show a map of Japan in a color code, forecasting which part of the country will soon see yellow leaves, reds, browns, and so on. In spring, this map appears again to mark the progress of cherry blossoms from south to north, reversing the process of the fall.

According to the Shinto code, the viewer on a proper leaf-viewing excursion should try to achieve a personal communion with the leaves, in a bond akin to the private communication between man and god at the heart of many Western religions. As Prince Genji once wrote to a lover, "A sheaf of autumn leaves admired in solitude is like damasks worn in the darkness of the night." By entering nature, one hopes to internalize the beauty of the leaves in one's heart. Man enters nature, and nature, in turn, enters man.

When one achieves this spiritual union, one is said to have learned the *wabi-sabi*: peaceful thoughts, peaceful action. This spiritual oasis is so fragile and difficult to attain that most young people do not even try. I learned this word not from Cho but from Kato-*sensei*, who, though nearing sixty, still regularly studied how to perform a proper tea ceremony. He lamented that young people no longer searched for such age-old dreams. Indeed, the few times I mentioned this term to younger friends, they laughed off my question as another example of the foreigner overzealous about things Japanese.

Still, I sensed an enduring legacy of *wabi-sabi* in the cycling and hiking of young people today. The bikers who had come to view leaves achieved that sense of escape into the outdoors, albeit with a little less respect for the sanctity of nature than their forebears showed. They had simply incorporated motorbikes into the modern "way of leaves." The same held true for us. Cho, Chieko, and I did not actively hunt the *wabi-sabi* when we climbed up tree trunks, swung from branches, and tossed leaves at one another, but unwittingly we had ended up on the same path as generations of leaf-viewers who had come before us.

After our day in the sunlight, Cho and Chieko and I returned to our classrooms, far removed from our momentary outdoor bliss. In time, the glow of Nikko began to dim. But then, as autumn approached its apogee, the wind picked up around Sano, blowing a torrent of twisters every day and stirring the branches of the chestnut trees. As promised, the wind was strong in Tochigi.

And the tempestuous Susano even delivered an unexpected gift with his gales. On a crisp November morning, as I rode across the narrow river on my way to Sano Junior High, the god of the Wind whisked away the clouds from the sky, and I spotted above the tile roofs and empty fields around town the familiar quiet face of Mount Fuji. The faint white mask of snow that shone above the hills was a reminder that even in Tochigi, deep in the hinterland, separated from the mountain by one of the largest, most modern cities in the world, we were still connected intimately to the most enduring symbol of ancient Japan. As Bashō wrote in 1676,

> *My souvenir from Edo*
> *Is the refreshing cold wind*
> *Of Mount Fuji*
> *I brought home on my fan.*

FROM BLACKENED SCALPS TO BOBBY SOCKS: THE ANATOMY OF A JUNIOR HIGH SCHOOL UNIFORM

Give every man thine ear, but few thy voice;
Take each man's censure, but reserve thy judgment.
Costly thy habit as thy purse can buy,
But not expressed in fancy; rich, not gaudy;
For the apparel oft proclaims the man.
— *Shakespeare*, Hamlet

THE EARLY MORNING FOG had not yet lifted when the students gathered on the baseball diamond in back of Sano Junior High. A mid-November chill hung in the air and elicited grumbles from the anxious students shivering in the cold. Boys and girls lined up according to class, with the ninth graders flanked by the underclassmen. A thin ninth-grade boy with black-rimmed glasses, who had been crowned the fastest runner in the *undō-kai*, stepped onto the pitcher's mound and began reading down a list.

"Cap. Who has forgotten their cap today?

"Scarf. Do all girls have their scarves?

"Gloves. Who is missing their gloves?"

As each item was called, the offenders squirmed in shame and meekly raised their hands while the teachers who patrolled the lines jotted down names on tiny notepads.

"Shirt. Whose tail is not tucked in?"

When the boy finished his inventory, he assumed his place in line, and the physical education instructor, Mr. Yamamoto, a square, well-built man with a salt-and-pepper crew cut and angry eyes, took the mound and began berating the students through a megaphone. How dare they not check their uniforms before they left home in the morning, he said. How could they expect to study in class if their jackets were not buttoned properly? How did they think their school looked to outsiders when so many had forgotten their nametags? At the end of his lashing, Yamamoto-*sensei*, the Bad Cop, stepped down and Sakamoto-*sensei*, the Good Cop, appeared.

"Attention," the sports teacher barked from the field, and a gentle shuffling ensued.

"Ready." The students faced forward.

"Bow."

"Autumn is here," the principal said with a smile, glancing from one row of students to the next and trying to assure them that he meant no harm. "Now is the season to take special care of ourselves. Now we must work harder to maintain our orderly world."

When he talked to the students the principal tended to speak indirectly, through metaphor. But sometimes he took action to reinforce his authority. Several weeks before this assembly, about the time of the annual change from summer to winter uniforms, three young boys had come to school with narrow parts shaved into their already shortened hair. The principal called the boys into his office, took a wide felt-tipped marker from his desk, and blackened the shaved portions of their scalps. The students knew that behind his veil of politesse the principal wielded a sword.

"Winter is coming," he continued. "The sky becomes darker every night. We must begin to adjust. Starting today we must all leave school by five in the afternoon, not six o'clock. Please remember to wear your jackets and bicycle helmets on your way home. You must be careful not to catch a cold. Our wind is very strong."

Sakamoto-*sensei* believed that clothes were a good indicator of character. He regularly monitored the shoe racks in front of the school to see which students were stepping on the heels of their sneakers instead of slipping them on all the way. This behavior, he said, was an early sign of delinquency. To drive his point home, he hung an old student uniform outside his office decorated with various warning signs: "DON'T SHORTEN YOUR LEGS"; "DON'T WEAR PLEATS"; "DON'T PUT PURPLE LININGS IN YOUR POCKETS." On the rope above the limp body, he had written his Shakespearian motto in script: "A CLEAN UNIFORM MEANS A CLEAN HEART." Finally, the principal required that all students carry a copy of the five-page school dress code with them at all times, in the shirt pocket across their hearts.

The first articles in this turquoise book covered clothes. All students, the code said, must wear the "standard-type" school uniform. For boys, this consisted of black pants and a tight-fitting black blazer with brass buttons and a high neck that scraped the chin like a clerical collar. Girls wore a matching navy blue skirt and blazer. Both boys and girls were required to wear a white shirt, but not just any white shirt would do. The code specified: "The white shirt must be pure white, with no wrinkles, no decoration, and no buttons on the collar." It must have a pocket big enough for a tag with the student's name written in large black letters. The code went on to say that the collars and cuffs of the shirt must be six centimeters long. In winter, students were allowed to wear sweaters over their shirts. Blue and gray were acceptable; pink and yellow were forbidden. V necks were allowed; turtlenecks were not.

The "standard-type" male jacket dates back over one hundred years to the Prussian military academies of the late nineteenth century. In the early months after the Meiji Restoration in 1868, a group of Japanese educators traveled to Europe to study pedagogical techniques and were so impressed by the discipline of the Prussian army that they adopted its austere military uniform as a model for their own schools. Despite all the changes in Japanese education over the last hundred years, the uniform has remained virtually unaltered. The only difference is that today the jackets are made of polyester, not wool.

In addition to the formal uniform, each student was required to own a set of "sports clothes." Boys' gym suits were royal blue, girls' were fluorescent orange, and both sported twin white racing stripes that ran down the sides. These clothes were used for physical education and other activities that required less formality. Each activity in the school day had an assigned uniform, and students changed between their formal wear and casual wear at least four times a day. It was not uncommon for students to come to school in formal wear, change to casual wear for lunch, back to formal wear for afternoon classes, casual wear for cleaning time, and finally formal wear for going home. To expedite these changes, students wore the T-shirts and shorts of their sports outfits underneath their formal uniforms.

Just as each activity required special outer wear, so each required special footwear. Every student was required to own at least four pairs of shoes: one pair of all-purpose sneakers to wear to and from school; one pair of slip-on shoes to wear in the classroom; a pair of slip-on shoes to wear in the gymnasium; and a specialized pair of shoes for club activities — cleats for soccer, hightop sneakers for basketball. In addition, the school distributed special plastic slippers for the bathroom and padded slippers to ease the anxiety of a visit to the principal's office. The code devoted several lines to describing the shoes that students were asked to wear to and from school. These shoes must have laces — no Velcro — and must have a "notched, mountainlike tread" suitable for negotiating potholes and slippery bridges. On these shoes, one brand mark or cartoon character was allowed as decoration; more than one was not.

Finally, every student must own a bookbag to tote his or

her belongings to school. This bookbag, the code insisted, must be carried on one's back, so as not to interfere with bike riding. Sportswear could be carried in a *sports* bag, but books must be carried in the *book*bag. To avoid any confusion on this issue, the code advised: "Even if you have other bags, it is ideal to bring all things in your bookbag. In fact, it is better not to bring many things at all when you come to school."

Although they may be masters of control, the Japanese did not invent the notion of discipline in school. Indeed, many of the regimental ideas used in Japanese schools could have been lifted line by line from the annals of Western philosophy. In the *Ethics*, Aristotle described the need to teach good working habits to young people through a strict regimen. "The mind of the pupil has to be prepared for the inculcation of good habits," he wrote a thousand years before the opening of the first Japanese school. "For this reason the nurture of young persons should be regulated by law, for hard conditions and sober living will cease to be painful when they have become habitual." The Japanese recognized the value of this Western tradition when they established their own modern school system in the late nineteenth century. The same expeditions to Europe which uncovered the German military uniforms returned with other ideas as well. From the French, the Japanese borrowed the idea of a centrally controlled authority, and from the British, they seized on the idea of moral education in school. They even adopted Western hair styles, encouraging all respectable men to cut off their old-style top-knots.

These Western imports were blended with more "native," Confucian ideas to make a Japanese ethic of discipline. The Confucian model contributed the notion of a well-ordered family controlled by a father, with service by his wife and loyalty from his children. This ideal of the proper family was reflected in school, where teachers, like fathers, dominated their students.

Confucianism is not a religion per se but rather an orthodox set of guidelines outlining how people should relate to one another, including the etiquette of conflict. Military skills were always taught in Confucian schools along with academic proficiency, and this emphasis on martial training became even more important in the early decades of the twentieth century. In 1939 Emperor Hirohito issued a call for all children to prepare themselves to defend their country. "We command you to put honor above all things," he said in the Imperial Rescript on Young People. "You should cultivate your literary power, learn military discipline, and muster the spirit of fortitude." This philosophy of combining selected Western institutions with traditional Eastern ideas has remained at the core of Japanese schools ever since.

Today, despite a brief period of liberalism following the Second World War, hard-line discipline is once again standard. The current policy marks a deliberate turn away from Western permissiveness, which many Japanese feel undermined their national culture. In the late 1960s, Japanese university students, like their counterparts around the world, erupted into riots against the government, demanding more individual influence over their education. These riots, bloodier than similar outbursts in the United States and Europe, became so intractable that in 1969 Prime Minister Eisaku Sato closed the prestigious Tokyo University for an entire year. Outraged that a group of dissident students could cripple the pinnacle of the nation's education system, the government cracked down on elementary and secondary students across the country to make sure that "alien Western notions" of individuality no longer polluted Japanese youth.

Sakamoto-*sensei* was a secondary school teacher during this time, and like many others he was disgusted that Japanese university students were letting their hair grow long, strapping riot helmets onto their heads, and storming their classrooms. Students were not supposed to attack their teachers, he thought, but respect them and follow their example. The solution, he concluded, was to pressure children when they were young to learn the value of discipline.

"Good schools have strict rules," he explained to me. "If the rules are stern and the students understand, then this school is okay."

But some of the younger teachers resented the way the principal imposed his stern will on the school. Machida-*sensei*, a handsome young math teacher who sat near my desk, told me that the principal had asked him to follow a *teachers'* dress code. "He said that I had to maintain a certain appearance. I should wear a tie but shouldn't put my hands in my pockets. He also said I shouldn't walk around the classroom but should remain behind the podium. I don't listen to him, so I don't expect my students to listen either."

Several days after the outdoor inspection a ninth-grade girl wore a pair of Mickey Mouse socks to school. Unsettled by this breach of conduct, the principal summoned all the teachers after school one Friday afternoon to discuss the problem of whether to permit students to wear socks decorated with commercial emblems. The meeting broke up after a short time, however, because the principal refused to accept the arguments in favor of name-brand socks.

"The other teachers say the principal is running the school too strictly," Denver confided in me after the meeting. "They say he is just like a 'one-man.' "

"Is that good or bad?" I asked.

"A 'one-man' is someone who doesn't listen to other people but just makes decisions on his own. In Japan, it is not a good thing." Denver looked around the room to make sure no one was listening. "He has a lot of authority and keeps everyone under control, but if something ever goes wrong in school he will have to take the blame himself. He alone is making the rules."

·　　·　　·

Although the regulations may seem relentless to an outsider, students still manage to create fashion trends within the limits they face. Fashion, however, is too important to be left to individual whim, so every year the ninth-grade students take matters into their own hands and devise a parallel, unwritten dress code that further regulates appearance, known in the halls as the *sempai*, or senior, code. The school code tells students what to wear; the *sempai* code tells them how to wear it.

Take socks, for example, which teachers thought so controversial. The school code stipulated that all students should wear white socks in winter but could wear mesh socks in summer. Yet mesh was considered such high fashion that the *sempai* code insisted that only ninth-grade students could wear these socks. The older girls also insisted that younger girls not fold down their tops in bobby-sock fashion, as this was deemed too suggestive for their age.

The *sempai* code varies from school to school and within each school according to the clubs. For example, while the color and style of each shirt was closely regulated by the school, the way to wear the shirt was left untouched. The upperclassmen at Sano Junior High leapt at this chance to teach their juniors, or *kōhai*, the value of respect. Seventh-grade girls in the volleyball club were required to wear their top buttons bound, while the ninth graders kept theirs unbuttoned. Seventh-grade boys in the karate club were ordered to keep their sweat suits zipped up to their chins, while the ninth-grade boys unzipped theirs to the waist.

The ninth-grade boys also declared that underclassmen could not make alterations to their uniforms, a common way of defying the code. Any seventh grader caught with a purple lining in his pockets or a red lining in his jacket would be punished — not by the principal but by a ninth-grade tribunal. Several months before I arrived, an eighth-grade student was caught with an embroidered dragon on the inside of his uniform. He was summarily taken to the bathroom by a band of ninth graders and beaten up. His jacket was confiscated and secretly passed around the ninth-grade floor.

But by far the biggest source of controversy in recent years has been the regulations governing hair. Among most students, hair is the measure of fashion. Many high school boys like to get "perms" or spike their hair with gel, while many girls tint their hair with orange or purple dye. To avoid this type of behavior in junior high school, the school code tried to be very explicit. Boys must have a *bōzu* cut, it said, meaning each hair could be no longer than five millimeters from the scalp. For girls, the code said the ideal type of cut was not too fashionable. "Girls' hair should be junior-high-school-like," it said. This directive caused enormous problems because the school did not indicate what "junior-high-school-like" meant or who would make this decision.

Naturally, the *sempai* code stepped in where the school code left off. Ninth graders declared that underclass girls must keep their hair above their shoulders and should not alter their hair color in any way: only straight black hair was acceptable. For girls in Tochigi, this was a particularly touchy subject. Japanese lore holds that only girls in the old capital of Kyoto have "pure" black hair, and the farther away a girl lives from Kyoto, the less black her hair becomes.

"The girls upstairs come down here during lunch," a seventh-grade girl griped to me during lunch one day. "They pull our hair and demand to know, 'Is your hair natural?'"

For those whose hair was truly off-color, this rule meant certain abuse. "My hair really is this color," said one girl with a reddish tint in her hair. "I think it's unfair that everybody's hair must be the same color. Coke changes color from can to can, so why can't hair change from head to head?"

For students, the message is clear: Conform. What the teachers do not patrol, the upperclassmen control, until every collar, cuff, and curl is covered by convention.

"You may think we are too severe," Sakamoto-*sensei* said to me after the emergency council, "but it's the little things that keep everyone in check. If we start to let the details slide, everything will get out of control. If we relax, all the students will become outlandish."

This is the maxim that guides school life and breeds the strict school code: If you give students an inch, they will take a mile; if you give them a choice, they might all dress up in designer socks or shave parts in their shortened hair.

8

INSIDE THE CIRCLE:
MAKING HOSPITAL ROUNDS

*Every city and town placed upon its porch, where it
could be seen by the eye of the Lady Moon, a tiny table
laden with treasure balls. There were rice dumplings,
chestnuts, and two circular sake vases. Everything had
been carefully selected as being the nearest a perfect
round in shape, for "round" is the symbol of perfection.*
— *Etsu Sugimoto*, A Daughter of the Samurai, *1926*

THE NOONTIME SUN gazed down from its crest and the wind
sliced through the sky as I crouched alone on the green clay court
awaiting the start of a tennis match against Mogi-*sensei*, one of
my fellow teachers from the Board of Education. For several
weeks the members of my office had placed bets on this match.
Surely Mogi-*sensei*'s overpowering forehand would prove too
strong for me, Mr. C insisted. Surely my serve — from its outlandish height — would sail right past his arms, Arai-*san* countered. Finally, on this mild November Saturday afternoon, we set
out to settle the score.

At first my size served me well, and I quickly jumped out
to a lead. But Mogi-*sensei*, a former science teacher with a passion
for sports, soon changed his strategy, employing wicked short
shots and topspin lobs to throw me off course. I was lunging for
one of his deadly dropshots late in the first set when I tumbled
head first over my knees and landed atop my left foot. Having
sprained my ankle in the past, I expected to rest about five minutes, get up, and continue with the match. I could not have been
more wrong.

Ten minutes later, without the aid of a dictionary and without the comfort of my mother, I was wheeled feet first into the emergency room of the Sano Kosei Hospital, with sirens blaring and a gaggle of nurses running by my side, rummaging through forms and spitting out comments in rapid Japanese.

"Wow, your toes are so *long*," one said.

"And these legs," said another, "so hairy."

Fearing that they would not stop at my lengthy toes and hairy legs, I managed a slight yelp of pain and was escorted down the hall to the x-ray room.

The black and white picture of my leg showed no splinters in the bone, but there was a small black pocket where white ligament should have been. Momentarily the doctor arrived, a young, disheveled man with heavy sideburns, tennis shoes, and a slight, worried smile. As he entered the room, the nurses took two steps back, bowed deeply, and asked for his gracious protection. Turning toward me, he began his diagnosis in Japanese. But after realizing I was having difficulty understanding his technical language, he took a deep breath and started again, this time in English.

"Ligament. Rupture. Cast . . . Shall we?"

He breathed an audible sigh of relief and gestured for the nurses to prepare the plaster.

"Excuse me," I interrupted, moving back into Japanese, "can't we discuss this a little more?"

Stunned, he sank back into a chair.

What followed was a rather arduous conversation as he repeated the same finding. "Ligament. Rupture. Cast . . . Understand?" I had the uneasy feeling that he knew about six words of English and was adjusting his diagnosis to fit his vocabulary. "You have cancer; we must amputate; have a nice day." Soon he stopped talking altogether, pulled on his plastic gloves, and declared, "Let's go."

"Yes, let's," the nurses cheered, lifting me to the table.

"But wait," I pleaded, "aren't there any other options?"

All the people around me — the doctor, nurses, and Mogi-*sensei* — were convinced that they should go ahead and wrap my leg in a cast. But I felt uneasy, mostly because the diagnosis had been so brief. (Later I learned that the Japanese have a special word for such diagnoses, *sanpun-kan shindan*, the three-minute treatment.) Should I submit to the will of everyone around me or follow my instincts and ask to speak to another doctor who I knew could speak English? Ironically, here I found myself inside the circle of a Japanese group, and all I wanted was to get out.

"Excuse me," I whispered to Mogi-*sensei*, opting for prudence over harmony, "perhaps we could call Dr. Endo."

Dr. Endo, the head of the Sano Public Health Department, was an elder statesman in the local medical community. I had met him on my first round of greetings in August and since that time had visited his home several times. He speaks fluent English, having lived and worked in the Philippines, Malaysia, and the United States, and has an affable charm that cuts through the formality of even the stiffest situation. But none of this seemed to matter at the time. To the people gathered in the hospital, Dr. Endo was first and foremost just another doctor.

"Why do you want to call Dr. Endo?" Mogi-*sensei* whispered back. "After all, this doctor is a bone specialist."

"But don't you agree . . . ," I said with a resolute smile.

During my early months in Sano, I had made the mistake of taking my problems directly to the person who I thought could fix them — my section chief or the director of my office. If I needed to change my schedule at school or arrange a meeting with my fellow teachers, I would issue my request in person and wait for an immediate response. But such direct communication rarely worked. Instead, I had to learn to make my wishes and opinions known more subtly — after hours at a bar with my boss, indirectly through the secretary, or simply up the chain of command. Like a seventh grader trapped by the strict rules of a *sempai*, I had to follow this code of hierarchy.

Having stated my case about a second opinion, I waited quietly for it to wend its way through the system. Mogi-*sensei* served as my proxy — the go-between in this arranged marriage — and relayed my request to the nurses. When the nurses informed the doctor, his face drooped and his eyes turned away. I felt my stomach sink. I had offended etiquette before in Japan, but never had I seen a faux pas register so vividly on someone's face.

Without meaning to, I had violated the sanctity of the teacher-pupil relationship by appearing to question the wisdom of the doctor. In most cases, the doctor decides and the patient accepts. Realizing that I had disturbed not only the doctor but also my hosts, I hastened to mend the rift. "In my country," I tried to explain, "we often seek the advice of two doctors." For good measure I added that Dr. Endo was not just any doctor but a friend of mine, and that I had been to his house for dinner just the previous week. Perhaps if they did not understand the need for a second opinion, they would see that Dr. Endo was part of my extended family.

A tense moment followed as they looked pleadingly at me to rescind the request, but I remained quiet, trying to convince myself that this was a time to withstand the pressure — to be Greek among Romans. Eventually the circle relaxed, and I was allowed to make the call.

If Japan had a national shape, it would surely be the circle. Not only does the circle appear on the national flag and on all currency (*en*, the symbol for money, actually means circle), but the circle comes closest to defining how the Japanese think about themselves. "We like round things," Mr. C once explained to me. "We want things to run smoothly with no sharp edges. Japan is an island nation, surrounded by seas and enemies, so we must depend on each other."

As an American in Japan, I continually felt a mix of wonder and respect at the way those around me made decisions, with one eye focused on themselves and the other on the groups to which they belonged. Although I met many strong-willed individuals, the majority of people I knew accepted the fact that their lives were controlled by the circles in which they lived and worked. That acceptance was not always wholehearted, of course: teachers complained of not being able to take Sundays off because they were forced to supervise volleyball practice at school; women told me that they wanted to continue working into their late twenties but were forced to retire at twenty-five when their bosses suggested it was time for them to marry; friends complained that they were not told when a parent developed cancer because the doctor felt the stigma of the disease would be too great for them to handle. As I lived in Japan, I struggled with this question: how much should I follow the unwritten rules that controlled the society around me, and how much should I remain attached to my own customs?

Many Japanese who have lived in the United States have experienced the same dilemma in reverse. Since Japan reluctantly opened its doors to the outside world over a century ago, a large number of people have ventured abroad to explore life in the West. Almost all who have written of this experience have expressed the same combination of surprise and unease at the "freedom" of life in America.

Takeo Doi, a psychologist and author of the 1972 best-seller on Japanese behavior called *The Anatomy of Dependence*, described a visit to a cocktail party in America at which the host asked him an interminable list of questions about his preference for a beverage. Would the guest prefer a soft drink or a cocktail? Would he prefer bourbon or Scotch? How much liquor would he like, and how would he like it prepared? "I soon realized," wrote Doi, "that this was an American's way of showing politeness to his guest, but in my own mind I had a strong feeling that

I could not care less. What a lot of trivial choices they were obliging one to make, and I sometimes felt as if they were doing it only to reassure themselves of their own freedom."

In Japan, the dream of personal freedom is not the Holy Grail that it is in the West. People aspire instead to drink from a cup that is well-worn and has first been passed around the room. Just as I struggled to accept the doting presence of the group in Japan, so my Japanese hosts struggled to appreciate the frustrating independence of this American. For the members of my office, this adjustment proved to be enlightening. Several weeks before I was impounded in the hospital, I was asked to make a trip to Tochigi's capital city of Utsunomiya to meet with other foreign English teachers. Mr. C came over to my desk at the Board of Education a few days before the meeting with a briefcase overflowing with maps and directions for making the one-hour train ride.

"Are you sure you understand?" he said with grave concern. "Do you need me to go with you?"

"I came from Georgia to Tochigi all by myself," I assured him. "I think I can make it safely from Sano to Utsunomiya."

Mr. C thought for a moment, then began to giggle like a boy. The whole office — ever listening — soon followed.

"Ah, he sure is an American," Mr. C said as he shuffled back to his desk. "He really has the Frontier Spirit."

Dr. Endo arrived promptly at the hospital, bowed to the teacher, used honorific speech toward the doctor, joked with the nurses, and generally loosened a tense situation. After conferring with the doctor he assured me that the cast was necessary. Then he smiled and relayed the doctor's recommendation that I be admitted to the hospital.

"The hospital," I gasped, "for a sprained ankle?"

"Yes," Dr. Endo answered. "You'll be unable to walk on your cast."

"Unable to walk?"

"Yes," the other doctor added, "but only for a week. Then I'll give you another cast."

Despite protestations, I knew I was trapped. Having placed all of my hopes with Dr. Endo, I could hardly have convinced my doctor and my colleagues that a third opinion was in order. Within seconds they had pulled a cotton sock above my knee and wrapped my leg in plaster strips. When they finished, I was allowed one telephone call, sent home, and told to return posthaste with a towel, a change of clothes, and a pair of chopsticks. This was a bring-your-own-flatware type of place.

Room 306 of the Kosei Hospital in midtown Sano looked like most other six-bed hospital rooms I had ever seen, complete with white tile floors, beige walls, and faded orange curtains across the window. When I arrived later that night, four other patients lay in various states of convalescence, with their wives and, in some cases, their children at their sides. When the excitement over my arrival subsided, I settled into a corner bed, and presently a nurse brought dinner. Pausing at the end of my bed, she looked first at me and then at the plate of rice and meatballs she held in her hand.

"Can you use chopsticks?" she asked.

I smiled at the question I had heard so many times.

"I sure can," I told her. "I even brought my own."

I had made it about halfway through the tray when another nurse appeared at the door, grabbed the tray, bade us good night, and turned off the lights. It was nine o'clock. For the rest of the evening I lay flat on my back with a lump of rice in my stomach, my left leg suspended in the air on a giant lime-green foam pyramid, and my head on a beanbag pillow. I spent my first night dreaming of escape.

Just as I wafted into fantasies of a daring breakout — catapulting to freedom from my third-floor cell or leading a platoon of one-footed convalescents — the curtains around my bed screamed open and a nurse appeared at the foot of my bed. "*Ohayō gozaimasu*, good morning," she chirped. "Did you have pleasant dreams?" I glanced over at the digital clock beside my bed: 5:55.

"Excuse me," I grunted. "I'm still sleeping."

"Oh no you're not," she chimed. "It's time to take your temperature." With that announcement she ripped aside my blanket, reached down my shirt, and stuck a cold thermometer underneath my arm. Then she promptly disappeared. I lay back on the bed and tried to resume my escape fantasies when suddenly the metal rod in my armpit started beeping like a smoke alarm. I jolted up, and the nurse came scampering over. "*Gomen, gomen*, sorry," she said as she reached into my shirt and extracted the high-tech digital thermometer. "Just right."

Then she pushed me back on the mattress, took my pulse, and started running through a list of questions in Japanese. I responded to each one in turn — no, I had not eaten all my dinner; yes, I had slept well — until she reached a final question, which I could not understand.

"Could you please repeat that," I said.

Off she went again. "Something, something, yesterday. Whatsit, whatsit, how many times?" Again, a blank stare from me. Not to be denied, she looked around the room, pulled the curtain closed, and carefully enunciated the word "TO-I-LE-TO," pointing to her rear end for clarification.

"Ah, got it," I announced. "Twice."

Then the nurse stripped back the covers again and pointed to my crotch. "Weeelll . . . ?"

Now fully awake and concerned that she had plans to probe me once more with her singing thermometer, I answered quickly, "The same."

Every two hours for the duration of my stay in the hospital, a nurse would appear in my room, stick one of those thermometers down my shirt, take my pulse, and ask me how many times

I had been to the toilet since my last checkup. Despite pleas that I was admitted for a twisted ankle and that perhaps we could relax this austere regimen, the nurses never slacked off.

Immediately I set about devising a strategy to gain my early release. First I made a list of Japanese expressions I could use to impress my visitors with the plight of my captivity. "Don't worry about me," I would tell every person who came to catch a glimpse. "I didn't break a bone. In fact, tomorrow I'll be going home." Soon everyone on the third floor of the Sano Kosei Hospital was discussing my injury and, more important, the good news that I would be going home the next day. If the fate of my stay was to be left to the will of the group, I planned to make certain that I had the will of as large a group as possible on my side.

The first day passed as I received visitors and collected a veritable magazine of gifts, including tangerines, plastic packets of Kleenex, and a bulging watermelon. But on my second day I learned that not I, not the nurses, not even the doctor, would decide when I was ready to leave the hospital: instead, my office elders would choose.

When Mr. C arrived early Monday, he made the prospects for an early release look dim. "Of course you can't go home now," he said. "Who would take care of you?" How pitiful I would be at home, he lamented, without a mother or a wife to tend to me. For safekeeping, it was best that I stay in the hospital for a while, where the "lovely and beautiful nurses" could dote over me. I knew I was doomed. All of the Kleenex packets and the tangerines in Tochigi could not dent the force of this logic.

Soon Mr. C was joined by Mogi-*sensei*, the man who had sent me tumbling in the first place, our section chief, and Kato-*sensei*, the director of the office. The four of them sat on the edge of my bed and talked about this grave situation as if I were not even in the same room. To them it seemed axiomatic that they would decide how long I should stay in the hospital. As a member of the office, and a young single member at that, I would naturally be dependent on them to make this decision. I would feel *amae*.

Amae, which can best be described as a feeling of dependence or reliance on others, has been called the core of the Japanese personality. In *The Anatomy of Dependence* Takeo Doi suggests that this sense of attachment is the glue that binds Japanese people to one another. Feelings of *amae* form the radii of the circles that link everyone in Japan in a network of interdependencies: children dependent on parents, parents on teachers, teachers on principals, principals on politicians, and politicians on the people. "By becoming one with the group," Doi writes, "the Japanese are able to display a strength beyond the scope of the individual."

Doi stresses that the group does not offer a substitute for individual identity but rather provides a context for assorted personalities to join together in pursuit of a common goal. Over time, I began to appreciate this view. Although I had originally felt an aversion to relinquishing control of my actions, I eventually learned not to resist the group so vehemently. None of these men had ever lived by himself. None of them had ever worked alone. Instead, all of them had come to see their colleagues as part of their *uchi*, or family. When they referred to themselves at work, they even used the word *uchi* in place of *watashi*, the familiar word for "I."

I realized in time that if I was to survive in their world — if I was ever to get out of this bed — I would have to satisfy my colleagues' sense of responsibility over me and at the same time maintain for myself the comforting myth of my own independence.

"Don't worry about sending me home," I assured them. "I will take good care of my leg."

"But how will you cook?" Kato-*sensei* asked.

"Well . . ." I paused for a second, then reached out in desperation. "I have friends, and they will bring me food."

As soon as I said this, I wondered if it was true. All my friends were busy with school; how could they bring me food at night? Kato-*sensei* looked skeptical. Mr. C tilted his head and sucked in his breath in consternation. Mogi-*sensei* nodded quietly. I felt the grip squeezing tighter, as it had in my inaugural bath. Then suddenly, as before, came relief.

Cho appeared at the door, carrying a bag of pears, and Denver stepped up behind him, toting a small TV set.

"We heard you were in the hospital," Cho said. "We thought you might need some help."

"You see," I said, beaming, "I have friends. One of them can bring me dinner on Tuesday, and the other Wednesday."

My office colleagues were startled by this turn of events but still gave little indication that they were willing to circumvent doctor's orders — until, to everyone's surprise, Dr. Endo appeared at the door with a coffee cake in hand.

What followed was true theater of the absurd.

Bow, bow.

"Oh thank you very much for your trouble."

Bow, deeper bow, bow.

"You took such good care of him."

Bow, bow. Step backward.

"Oh no, not me."

"Oh yes, just you."

Bow.

Nod.

"He's so far away from home."

The other patients marveled at this motley crew assembled in Room 306: four members of the local education office, two young teachers, one doctor with a coffee cake, and a six-foot foreigner with an engorged leg and a year's supply of Kleenex.

Dr. Endo sized up the problem immediately and signaled a breakthrough by suggesting that since I would be unable to prepare food for myself, perhaps I could use crutches, or *matzubazue*, to leave the house and eat. I watched as the idea was nodded up the chain of command, from my doctor to my boss, to our section chief, and finally to the approving glance of the director of our office. Just like that, I was free to go. All I had to do was clear two minor hurdles: first, get the doctor's approval; and second, find a pair of *matzubazue*, or "pine branches," that would work for me. The first task was easy, as the doctor had no choice but to relent when faced with the consensus of all my protectors, but the second proved difficult, as the nurses ran into unexpected trouble finding a pine tree with branches long enough to fit my frame.

Giant trees are not unknown in Japanese lore. Legend holds that the Sun goddess Amaterasu once planted a pair of chopsticks in the ground near Kyoto and that these sticks later grew into towering Japanese pines. In another instance, a Buddhist priest exhausted from a long pilgrimage planted two chopsticks in the ground to pray for a safe trip home, and these blossomed into giant willow trees. (The moral of these fables is that chopsticks do not grow *on* trees; they grow *into* trees.)

The first pair of crutches the nurse found were antique and wooden and reached about as high as my navel. The next pair, when extended to full length, actually reached my shoulders, but the handles remained at my knees. I had begun to fear that we would have to send to Tokyo for a suitable set when a nurse came running in with a new pair of pine branches made of aluminum. Everyone pulled in close to watch as I suspended myself from the handles, lifted my foot from the floor, and — behold — did not tumble to the ground. At last I was free to go.

Hurriedly I packed my bag and collected my towel, clothes, and chopsticks. I dispersed the various cakes, comic books, and packets of Kleenex to the other patients, and set out with my entourage. As I shuffled down the hall toward the elevator, already learning to bow from atop my branches, the head nurse came rushing out of her glassed-in office. By now, this woman knew me well. She had heard me speak Japanese and seen me eat rice; she knew exactly how many times I had been to the bathroom in the previous forty-eight hours. But the foreigner in Japan never seems to lose his ability to surprise. The nurse came to a sudden stop in front of the elevator, gaped at my leg, gazed at me standing there in front of her, and in an earnest, concerned tone asked, "Can you use crutches?"

We had come full circle. "I sure can," I told her. "After all, they're just like chopsticks."

9

TRASH DAY:
PLEDGING ALLEGIANCE
IN JAPANESE SCHOOLS

*In the administration of all schools, it must be kept in
mind that what is to be done is not for the sake of the
pupils, but for the sake of the country.*
— *Mori Arinori, Japan's first Minister of Education,
1885*

THE FORTY-FIVE STUDENTS in the first homeroom class of
the ninth grade were all seated at their desks when the opening
notes of the Brahms symphony roared from the loudspeaker at
precisely 8:30 A.M. Soon the violins faded and a slow, synthesized pulse spread across the room, numbing the mind with its
smooth, hypnotic gait. The room was cold and slightly dank. No
sun shone through the plate glass panes overlooking the balcony.
The clouds, like the students, were still.

In a moment, a soothing, resonant voice began to speak.
"Good morning, boys and girls. Let's begin another *wonderful*
day. Please close your eyes . . ."

For ten minutes every morning the students at Sano Junior
High sat in quiet meditation to prepare themselves for the day
ahead. The principal, Sakamoto-*sensei*, had introduced this system, known as Method Training, several years earlier in an attempt to quell the growing incidence of school "violence," mainly
minor scuffles and hair violations. The program consisted of a
sequence of twenty-five tapes for total mental and physical con-
ditioning. Each day a different tape was played.

After a pause, the breathy voice returned. "Concentrate on

relaxing your body. Allow your right arm to hang loose by your side. Focus your energy on that arm. You feel a strong, heavy sensation. Deep . . . Heavy . . ."

As the students followed the invisible commands, Mrs. Negishi, the homeroom teacher for this class, wandered among the rows of desks and monitored the appearance of each child. As she walked, she jotted comments on a small notepad in her hand. "The ones who open their eyes or who totter in their chairs are the poor students," she confided to me. "Those who do not concentrate now will not be able to work well later in the day."

On this day especially, students would have to cooperate with one another. They would attend regular classes in the morning, then take the afternoon off to join in the annual prewinter ritual known affectionately as *gomi-no-hi*, Trash Day. Working with their classmates, students would file out into the alleys and vacant lots around town and spend several hours picking up litter. Because of the importance of this day, Mrs. Negishi asked if I would like to join her class. Although my leg was still encased in a walking cast — the crutches were long since gone — I readily agreed.

After ten minutes the music dissolved, the voice disappeared, and Mrs. Negishi — standing erect before the class — took control of the homeroom meeting.

"Stand up," she commanded, and the students rose to their feet.

"Attention," she said, and they dropped their arms to their thighs.

"Bow."

It was 8:42 in the morning.

In every school across the country, students are assigned to a homeroom class, or *kumi*. The word *kumi*, which is rooted in the Japanese character for thread, was first used several hundred years ago to describe small bands of samurai warriors attached to a feudal lord. Like these faithful fighters of the past, students learn today that duty and honor begin with dedication to this group.

The first *kumi* of the ninth grade, under Mrs. Negishi, epitomized the intensity of this group affiliation. The forty-five students of the "9-1" class remained in the same room, at the same desks, for seven fifty-minute periods a day, five and a half days a week, forty-five weeks a year, with only brief escapes each day for physical education and science lab. Because the government allows no tracking of students based on ability, the members of this class reflected a true cross section of the west side of Sano. Future scientists learned alongside future truckdrivers, future poets along with future store clerks. While this system presents countless problems for teachers, who at any given time are speaking over the heads of some students and under the heads of others, the government feels the advantages for social relations are more important. The future doctor learns early to give assistance to those who are less capable.

Because they spent so much time together, the students in this class had developed an intimate bond with one another, not unlike the "relationship without clothes on" that office workers strive to achieve with the ritual bath. The students meditated in their chairs in the morning, ate lunch at their desks at noon, and changed clothes in the middle of their room at least several times a day. Like most tight-knit teams, the students teased one another inside the locker room, but outside of class they maintained a common front. During recess after lunch, for example, most baseball games were based on homeroom loyalties. The boys chose teams; the girls cheered; and other classes were not invited. Once, as I arranged to take a picture of one of these games, the leaders of the class shooed away stray fans from a neighboring *kumi*: 9-1 students only, they insisted.

The importance of the *kumi* as an educational tool, however, goes beyond this good-natured team spirit. The classroom

that the students inhabited had been, in effect, leased to them by
the school, in an arrangement not unlike the way a feudal lord
lent land to a group of serfs. By taking possession of this plot, the
students were able to practice tending their own home, cultivating their own garden. They made posters of their class motto to
hang on the wall; they kept plants on the balcony rail; and they
sometimes brought flower arrangements from home to put on
the cabinet in the corner. Every morning before school, students
rummaged around the room, sponging down the blackboard,
replenishing the supply of chalk, and writing the day's schedule
on the board. The chores changed on a rotating schedule and
gave each student a chance to practice "preparing the farm" for
a day.

At 2:30 every weekday afternoon and at noon on Saturdays,
classes officially ended and the daily ritual of cleanup began. The
students changed from their formal dark uniforms into their
colorful sweat suits, covered their heads with white kerchiefs,
and dispersed into small groups to tidy the room. As Madonna
or some other pop star blared on the loudspeaker, students
stacked desks and chairs into a corner, soaped down the blackboard, and threw out the trash. Boys cleaned the windows and
girls vacuumed the erasers (every classroom in the school had its
own electric, dust-sucking machine that automatically inhaled
residue from erasers). Cleaning time, like lunchtime and homeroom before it, was often frenetic and fun. A group of boys
would stop to arm-wrestle while some girls arranged a tournament to find out who could crawl the fastest across the floor with
a dampened rag. It was hard work, but as one student said, "If
the room is clean, we like to study more."

Although many teachers resented having to mop the floor
when they had more important work to do, they still viewed
their role as vital. Mrs. Negishi also changed out of her skirt and
into a sweat suit (although not in the classroom), wrapped a
kerchief around her head, and scrubbed the floor alongside her

students. She taught by example. "If I don't clean, the students don't clean," she told me. "It's part of my responsibility." This is one of the main tenets of the *kumi* system: a partnership between students and teachers. Together they work to promote the welfare of the community and foster the hygiene of its members. Students and teachers have clear roles, but the success of each depends on the cooperation of the other. At times the teacher plays disciplinarian, at times counselor, at times trusted friend. It is no wonder that students who grow up in this nurturing and protective environment learn to be dependent on those around them for everything from answering questions in English to deciding when to leave the hospital. From here students need only make a short leap of faith to transfer their trust in the *kumi* to reliance on their corporate co-workers later in life.

Beyond their commitment to the homeroom, however, students also learn to be aware of the higher structure that allows their *kumi* to prosper. Once a week students deferred tidying their own rooms to clean the entire school. They emptied the ashtrays in the teachers' room, scrubbed the toilets in the bathrooms, and pulled weeds from the garden in the parking lot. Before the *undō-kai* in October, students even clipped the grass on the playing field with classroom scissors. But the ultimate lesson for students is that they must look beyond their homeroom and their schoolyard to fulfill their obligation to the community as a whole. For this purpose, schools developed Trash Day.

"EVERY DAY WE ARE AWAKENING TO OUR OWN NEIGHBORHOOD," screamed the headline atop the student handout. "LET'S FRESHEN OUR CITY TODAY." At an outdoor rally just after lunch, each student received a mimeographed map, marked with a route for his or her *kumi* to follow, and two empty plastic bags: one for paper, the other for aluminum cans. After the guidelines had been explained, the principal took to the pitcher's mound and reiterated the theme of progress.

"This year we want to press toward greater cleanliness," he yelled. "This year we want to achieve a new order. Let's go out and make our city proud!"

After the pep talk, the members of 9-1 headed west out the back gate, toward the span of factories and small plants that lined the outer fields of Sano. Like other events during the school day, this seemingly burdensome activity was carried out with great verve and enthusiasm. The students raced in and out of muddy ditches as well as up and down trees in search of unsightly debris. The local newspaper sent a photographer; shopkeepers and mill workers emerged from their buildings to cheer the students on. The whole experience felt like a holiday parade. My students were so spirited that they even carried their class flag on the hunt, a red and white banner with a caricature of Mrs. Negishi, emblazoned with the slogan "9-1 IS NUMBER 1."

As I limped along with the students, I decided to teach them an American game, I Spy.

"I spy something red," I would cry, and the students raced to retrieve the prize. Eventually I began giving points for every item I spied that students named in English.

"A box."

"One point."

"A bottle."

"One point."

"A potato chip bag."

"Two points."

This little diversion occupied the students for most of an hour, especially after they realized that I would give extra points for complete sentences. "This is a pen" earned four, and "I see a tire" earned five. But the biggest awards of the day went to objects so extraordinary that they needed no verb. I gave a ten-point bonus for a Georgia Coffee can and a five-point *penalty* for an "adult magazine." This phrase was Living English, all right, but not what the government had in mind.

Compared with the normal drone of classes, events like this

were thrilling for the students, and they reinforced the message that community service can be fun when performed in a group. Trash Day was a painless way to teach students that their rights as students go hand in hand with their responsibilities to the nation. The only problem in the course of the afternoon was that the students felt they had not met this year's goal of greater cleanliness.

"I was embarrassed by how much trash we found this year," one of the girls in 9-1 complained at the end of the day as she dumped her cans into a recycling bin. "Our city should be ashamed."

"Our school should be ashamed as well," her friend added. "The streets around here are just a mess. I think we should ask the PTA to help us with this problem."

"But what can they do?" the first girl asked.

"Maybe they can go with us," her friend answered. "We could pick up trash together."

The first girl thought about this idea and agreed that it was worth a try. "If our parents don't care about this problem," she declared, "then we should show them how."

Japan has no pledge of allegiance. Most classrooms display no flag, and the national anthem — a wistful paean to the imperial line — is rarely played in schools. Yet Japanese schools succeed in teaching students a profound and lasting national pride. From Method Training every morning to Trash Day once a term, students learn the importance of working with a group and serving their community. In short, they learn to be good citizens.

In terms of time, Japanese students spend twenty-five percent more days in school than Americans, so a high school graduate in Japan has spent as much time in class as a college graduate in the United States. In terms of achievement, Japanese students consistently outperform their international peers in math and science achievement tests. In terms of dropout rates, ninety-five percent of Japanese children graduate from high school, compared with seventy-five percent of Americans.

But beyond these statistics, Japanese schools succeed on a more profound level of preparing their students to become productive members of society. In Sano, as elsewhere, schools are the focal points of neighborhoods. Students cycling through town are treated with respect. Teachers hold esteem in the eyes of the community. National news media give extensive coverage to regular school events, like the sports festival, the entrance exams, and the changing of school uniforms in October and June. All this attention serves as a constant reminder that education is vital to the continued prosperity of the country. Schools are successful in Japan for this simple reason: they are seen as a national security priority. Most Japanese know that their country has few natural resources — no oil, few minerals, limited arable land — so they learn to exploit the one resource they have in abundance: people. Beginning on their first day in school, students learn a familiar refrain about their country: "Japan is a small island nation with few natural resources, which is surrounded by countries that are bigger and stronger and out to weaken us. If we are to succeed, we Japanese must work harder and longer to overcome these odds." In essence, this has become the Japanese pledge. By stressing this code and encouraging children to sacrifice their personal desires for the good of the country, schools have been able to achieve what is, perhaps, their highest calling: to forge allegiance to the state.

Activities such as cleaning the school and clearing the neighborhood of trash are part of the Ministry of Education's overall plan to encourage a national identity. As expressed in the *Course of Study*, the government hopes such events will help produce citizens who will "love our nation and strive for our nation's advancement on the one hand, and contribute to the welfare of mankind on the other." While these words are lofty, as are those in our Pledge of Allegiance, they seem to hold meaning in the daily lives of students — perhaps more so than the words of our pledge, "liberty and justice for all."

The exhaustive emphasis on group training in Japan also has negative side effects, especially on students who for one reason or another feel left out of their *kumi*. In my early months as a teacher, no one mentioned to me that students who live abroad for a while are often shunned by their classmates when they return to Japan. No one told me that the country still suffers from the legacy of a four-hundred-year-old feudal class system that was officially outlawed over a century ago. And no one warned me that certain students are ostracized by their peers because they come from families that are still tainted by this past. In Sano, all of these problems would boil to the surface in the course of my year as a teacher, and one would end in tragedy.

While the *kumi* network has definite drawbacks, the system triumphs in one of its primary goals: to develop a community ethic among most students. Through repetition and eventually habit, students learn that they should spend a part of their day, indeed a part of themselves, tending the world around them. What begins in the homeroom at school later becomes the spirit of cooperation in many companies which so many Westerners admire. The lesson from the *kumi* is that this spirit is not mysteriously passed down through management seminars or religious rituals but is systematically and deliberately taught in schools. For students who pass through this system, a simple axiom serves as their personal pledge of allegiance: This above all, to thy *kumi* be true.

About two weeks after Trash Day, the students at Sano Junior High published the second-term edition of their school newspaper. One page of the four-page "Sano *Shimbun*" was devoted to the problem of too much trash.

"ARE WE PROGRESSING TOWARD CLEANLINESS?" the headline asked.

The answer, according to a questionnaire distributed by the newspaper staff, was a resounding no. Seventy-two percent of the students said that the amount of trash had stayed the same or increased from the previous year. When asked what they proposed to do about the problem, fifty-seven percent agreed that the school should sponsor more Trash Days. Another third recommended going door to door and asking for cooperation, and a final group suggested that every student carry a bag at all times and pick up trash on the way home from school.

In a box in the bottom right-hand corner of the Trash Day page, the newspaper staff printed two personal testimonials under the headline "MEMORI GOMI." In the spirit of the new crusade, the leading item was a "Good Memory" from the ninth-grade student Kyoko Susumu:

> It seems that up until now we tossed garbage without really thinking about it. But starting with the teachers and adding the help of other community members, we are attacking this problem with a great deal of enthusiasm.
>
> When we pick up trash, we grow both mentally and physically. At the same time we make our town look beautiful. From here on out, I would like to see the mind-set that holds no connection between throwing things out and the beauty of our town change to one of picking things up.

For equal time, the staff ran a "Bad Memory" from the eighth-grade student Koichi Nakamura. Even in his negative thoughts, Koichi sought to rally the school behind the banner of change:

> My negative thoughts about the garbage are that there are piles of trash right next to our school routes and there is hardly any room left to walk. Local factories put their big, bulky garbage here, and people passing by toss their trash on top. This is really disgraceful.

There are many places like this around town, and I think everybody in this area needs to cooperate. The most important thing is to be conscious of throwing things out in designated places. If every single person were to follow this rule, I think we could eliminate these "bad memories."

In its final article the newspaper reported that the growing problem of trash and the results of the schoolwide survey had been discussed at the most recent meeting of the Sano Junior High School PTA. When faced with the overwhelming evidence gathered by the students, the parents and teachers agreed that the situation was dire. Following the recommendations from the student council, the PTA agreed that the students should post signs around town, distribute leaflets, and even hold an additional Trash Day the following term on a Saturday afternoon so that parents could join in the hunt.

With all of this activity focused on the problem, the newspaper staff concluded that a "new age" of tidiness was about to dawn. They closed their coverage of Trash Day with this manifesto:

With a new tie-up between the students and the parents, we can soon erase this problem. On the day that we no longer need to hold any "Trash Exercises" we will know that our circle of cleanliness has spread throughout the town. Why don't we work toward this?

10

BOTTLED MILK AND PLASTIC CHOPSTICKS: THE LOST ART OF SCHOOL LUNCH

There is no doubt that the conveniences of modern life, by taking the place of human hands and feet, have robbed our children of the tools they need to succeed.
— *Daisaku Ikeda,* Glass Children, *1983*

AT THE END OF THE FOURTH PERIOD on a cold December morning, an eight-tone chime echoed through the corridors of Sano Junior High. At the tone, time seemed to stop for a moment; thoughts were halted; ears turned to the speakers on the wall. Then the bell sounded a second time, and the silence succumbed to a growing clamor of shuffled papers and screeching desks that ricocheted through the halls. School lunch had begun.

Within seconds, teachers came pouring into the office where I sat preparing for an afternoon class, and students came sliding in after them, lugging stacks of books and carping about the cold. The assistant principal looked up from his desk and smiled at the frenzied scene. A moment later two young boys tiptoed into the room, looked quietly from side to side, and stopped just shy of my desk. Their blue sports suits were zipped to the chin; their name tags announced their class: 7-4. The boys stared at their hands for a second; then the taller of the two whacked his friend on the head and pushed him in the back, saying, "Go ahead, ask him."

The boy caught himself on the lip of my desk, stared at my face like a frightened deer frozen by headlights, and then pushed himself erect again and addressed me face to face: "Mr. Bruce. Come our class. Lunch?"

The other teachers smiled; the boy looked pleased. I began to gather my papers into a pile when suddenly the first boy rushed forward, slapped his friend again on the head, and shouted, "Hey, stupid, you forgot to say 'Please.'"

"Oh no," the boy cried, "I make a mistake." He slapped his knees and turned back toward me. "Mr. Bruce. Come our class. Please?"

The teachers applauded, and the young solicitor drew his hands from his pockets and pushed his friend out the door. "There," he said. "I did it."

In the hall, dozens of students rushed around with great intent, shouting orders to one another and hurrying along in small platoons. It looked at first as if a student-run coup d'état was in progress. Sano Junior High, like all others in Japan, has no central cafeteria. Instead, students ate lunch in their classrooms. At the sound of the twin chimes every day, each *kumi* separated into predetermined teams to prepare for the midday meal. One team from each homeroom raced to the kitchen on the first floor and returned lugging huge pots and plastic bins containing the day's supply of salad, stew, and rice; another group rearranged the room by pushing the desks into six tablelike formations; another donned white aprons and linen kerchiefs and prepared to dish out the food to the class. The idea of having the students arrange and serve lunch every day, then clean up afterward, clearly saved money for the school, but the primary reason was for students to practice working together in a noncurricular activity.

Every day I ate with a different group of students; this day I would spend with Denver's homeroom class. Although Denver was only what he called a freshman teacher, he assumed the role of my elder in school. He would explain minor rules, interpret fights between students, and inform me of any points of history he thought I should know. In the same way he took care of me after I got out of the hospital. He brought me not only a television set but also a VCR, some tapes, and some English-language books, including Ezra Vogel's *Japan as Number One*, which he had been assigned to read in college for his English debating society. Like Cho, Denver had attended high school in Sano and then gone to Tokyo to attend a private university. After graduation Denver worked for several years at a bank in Tokyo before returning to Tochigi to become a teacher. In his spare time he still read business gossip magazines and toyed endlessly with his computer. Every night after returning from work, he would sit in his tiny second-floor room at his parents' home, smoke cigarettes, drink orange soda, and enter grades and comments about each of his homeroom students into his NEC personal computer.

Denver, like all other teachers, joined his homeroom students in their classroom for lunch every day. While this might seem a sure-fire formula for a burdensome hour, the mood was quite relaxed. Lunch hour was lively, at times raucous, and almost always jovial. Denver loosened his tie, changed from his black blazer into a fire-engine-red sweat suit, and seemed to savor the change in tone from the morning classes.

"Hey, *sensei*," one girl shouted from the back of the room, "got a girlfriend yet?"

"Not yet." He shook his head.

"Guess not," cried another. "You're getting too fat."

When the lunch pots arrived, I took my place in line with the students. The menu for the day consisted of steamed white rice, salted cucumbers, oranges, and a creamy stew with sautéed beef, carrots, and *konnyaku*, a rubbery gray substance made from the root of a Chinese tree and nicknamed "devil's tongue" by the Japanese. After receiving my food, I was handed a single metal eating utensil that was shaped like a spoon but had tines like a fork. "This is a spork," the girl explained.

Denver led me to the back of the room and a seat across from him. As I squeezed into the chair, stretching my legs around the outside of the desk because they refused to fit underneath, Denver surveyed the meal.

"This is a typical Japanese lunch," he said with a satisfied look. "Do you know how to eat it?"

I thought this was a joke about chopsticks until I recalled that we had sporks instead.

"Sure," I said with deep sarcasm. "Take the spork and eat some rice. Then take the spork and taste some stew. Then take the spork and eat some salad. Just as we do in America."

"Oh no," he protested. "You cannot eat this meal as you do in America. That would be a mistake. You have to use the *sankaku-shiki tabekata*."

"The what?"

"The *sankaku-shiki tabekata*," he repeated. "The triangle eating style."

By this point all the students had been served, and Denver stood up to call the class to order. The students settled into their seats and drew their hands together as in prayer.

"Ready . . . ," Denver called. "Begin."

"*Itadakimasu*," the class said in unison, repeating the simple expression of thanks that is said before all meals in Japan.

"Now I will teach you the proper way to eat Japanese food," Denver said with characteristic enthusiasm as he returned to his seat. "First, take a little rice and put it in your mouth. Then take some meat, and finally some salad."

I followed his instructions as he spoke.

"Now chew them all together at once. Rice. Meat. Salad. These are the three sides of the triangle."

Soon all of the students at our table of desks were demonstrating the proper technique. "One, two, three, chew. One, two, three, mix."

"You see, rice has no taste," he explained. "So it is best to take other food and mix it with the rice. This is why we don't use plates at home. Each person holds a small bowl of rice while we place several serving plates of food in the center of the table. We pick up the food with our chopsticks, bounce it on our rice, and then put it in our mouths. This way the food tastes more delicious, and it's better for our health."

"How can putting all of that food into your mouth at one time be good for your health?" I asked.

Some of the students giggled at the question.

"It's not *how* you eat," he said, "it's *what* you eat."

By now most of the students had finished their meal and were straining to hear our conversation.

"In Japan, we have certain eating habits," he continued. "When we are young — like these students — we eat a lot of meat. It's good for us while we're growing up. But when we get a little older — like I am — we begin to eat more and more traditional Japanese food, such as mountain vegetables and miso soup [a plain broth made from soybean paste]. By the time we get older — like the principal — we eat *only* Japanese-style food."

"It sounds like a diet triangle," I said, sporking a last mouthful of rice, stew, and salad and swishing them like mouthwash from side to side.

"Americans only eat meat," he said. "Children, parents, even grandparents eat meat. I hear you even eat meat for breakfast. I think this is not very healthy. You should try the Japanese way. Then you would live longer."

"*Sensei, sensei*," a young girl said from the front of the class. "Look at the time."

Denver glanced around the room and realized that we were the only ones still eating. Standing on his chair, he called the class to order and thanked the students for their patience.

"Are you ready?" he said, and the students again grew quiet and raised their hands to pray. "Okay."

"*Gochisō-sama deshita*. Thank you very much for the treat."

· · ·

Nothing symbolizes Japan's shifting attitude toward the West in the last 150 years more than its views on beef. For most of its history, Japan was a land of vegetarians. Buddhist taboos against eating animals and the scarcity of game in general meant that red meat was considered a foreign food. But with the coming of outlanders in the mid-nineteenth century, red meat quickly became a status symbol, and restaurants specializing in beef dishes popped up in major cities. Among the new urbane class, a man was not considered civilized unless he strapped a Western watch on his wrist, sprayed himself with eau de cologne, and dined on *gyūnabe*, beef stew.

Red meat was the first of many foreign foods to capture the hearts of Japanese youth and make its way into the national diet. By the early twentieth century Japanese people were growing fat on American beef and getting drunk on European whiskey. But in the 1930s and 1940s wartime austerity and trade restrictions brought these indulgences to an end. During the height of the Pacific War, when Japan was almost completely cut off from the rest of the world, the Japanese diet dwindled to its lowest level in terms of calories in nearly a century. In 1945 the average daily intake of a Japanese adult consisted of only 1,200 calories, less than is found in an average junior high school lunch today. Older Japanese still talk about those months as a time of overwhelming hunger. The American commanders who entered Japan in August 1945 quickly realized the severity of the situation and overlooked their wartime hostilities to airlift food into the country. "Send me food," General Douglas MacArthur said, "or send me bullets." This single act by the Americans toward their erstwhile enemies did as much as anything else to secure friendly postwar relations between the two countries.

In addition to bringing in short-term supplies, the occupying Americans also introduced a nutrition program into the schools, promising every child one glass of American-made powdered milk every day. After the Americans left in the 1950s, the

Japanese passed a school lunch law of their own which guaranteed that students would have one meal every day. As a sign of increased prosperity and friendship with the West, beef was included in the official menu. This program has been so successful that today it serves over sixteen million lunches a day.

I first realized the importance of school lunch when I was a patient in the hospital. Dr. Endo explained that the menu in the hospital would be good for me because it was modeled after the one used in schools. This menu, he explained, which includes milk, beef, and other ingredients not part of the traditional Japanese vegetarian diet, was the single most significant factor in improving the physical make-up of the population. Japanese children have grown dramatically taller in the last half a century. In 1950 the average fourteen-year-old boy was four feet, ten inches tall. In 1988 a boy the same age had surged half a foot to five feet, four inches tall. The average girl has grown from four feet nine to five feet two. Students weigh more as well. The male student today weighs twenty-eight pounds more than his counterpart of 1950.

While the typical school lunch menu has come a long way from its meager beginnings — now including such delicacies as shredded beef, shrimp stew, and fried squid — the program has not been without controversy. Menus in recent years have been criticized as being too high in calories and fat. As a result, national guidelines now limit fat to no more than thirty percent of total caloric content and have set new goals for protein, vitamins, and minerals. These eating patterns, combined with a general tendency to eat less red meat than in other industrialized countries, ensure that a Japanese junior high student can expect to outlive his or her peers in every other country in the world. A girl born today can expect to live to be eighty-two years old and a boy nearly seventy-six.

Another debate has centered on the question of what the staple of the meal should be. Rice has long been the main ingredient of the Japanese diet, and until recently it was eaten three times a day. *Gohan*, the word for cooked rice, is also the word for meal. But because of rice shortages after the war, bread was introduced into school lunches and soon became the norm. After twenty-five years of sandwiches, buns, and cinnamon rolls, the country's rice lobby complained. As a result, schools in the 1970s began serving rice twice a week from the huge store of surpluses that the government maintained in order to subsidize farmers.

In the meantime, students had gulped down so many doughnuts and hamburgers that they had forgotten the right way to eat rice. Parents noticed with horror that their children were not using chopsticks, or *hashi*, with the proper technique. Errant youngsters were shoveling their rice, stabbing their meat, and crossing the tips of their sticks in an X, the culinary equivalent of walking bowlegged. Like other lapses in manners, this one was blamed on the schools. In the late 1970s the Japanese Ministry of Education conducted a survey and found that less than ten percent of all schools were providing their students with *hashi*. Suddenly chopsticks, those little tapered slivers of wood adapted from Chinese ivory prototypes over a millennium ago, found themselves at the forefront of an all-out national effort to reclaim the lost art of eating. The cultural push to preserve good eating habits coincided with the political drive to bring back rice to the schools, and the result was a new national slogan of sorts: "A pair of chopsticks in every hand; a bowl of rice on every desk."

By the late 1980s this trend seemed to be holding, and ninety percent of all schools reported that they used chopsticks at lunch. But this study must have ignored the students in Tochigi. In Sano, chopsticks were the exception, not the rule, and students were clearly more comfortable using their hands or their sporks. Part of the problem for children may be that schools distribute short, stunted sticks made from cheap plastic. Some experts claim that Japanese children have actually outgrown traditional *hashi*. A study by a university professor in Tokyo concluded that the length of *hashi* should properly be about fifteen percent of a person's height. But with typical classroom chopsticks only seven inches long, the contemporary model had failed to keep up with the rapid growth of its target constituency. Using this formula as a base, the boy from the seventh grade who invited me to lunch would be more comfortable with sticks that were nine inches long, almost twenty-five percent longer than the ones he was using. Denver would need ones that were ten inches long; and I, the lofty foreigner who had already stunned so many with his chopstick proficiency, should rightfully have been using sticks that were similar in length to an American classroom ruler or, perhaps, a Japanese hospital crutch.

After finishing their beef stew and wiping their bowls with a tissue, the students of 7-4 hurriedly stacked their trays, deposited their orange peels in the trash, and put their milk bottles in racks. As they returned to their seats, Denver took his place behind the teacher's desk.

"Well, boys and girls," he said, leaning forward and scanning the room with his eyes, "how were classes this morning?"

Nobody answered. A stack of bowls fell to the side of the metal milk crate.

"How about sports class . . . Was it cold?"

Several students nodded, but still no one spoke out.

"Math class?"

"It's too hard," a boy in the back of the class whispered under his breath, just loud enough for everyone to hear.

"What did you say, Doi-*san*?" Denver called out. "Speak up so we can all hear."

A young girl in the front row raised her hand. "*Sensei*, tomorrow we have a test in both math and Japanese. Two tests in one day. That's not fair. All we do is study these days."

After lunch every day, Denver held a brief meeting with his homeroom students to discuss their classes and their behavior that morning. These noontime sessions, like spiritual revivals, were designed to renew the students' faith in their *kumi* and strengthen their commitment to one another. The meetings also served as an initiation into conflict resolution, the foundation of Japanese law. On this day, after a slow start, he calmed anxieties about upcoming tests and, more important, uncovered a minor class scuffle.

"*Sensei*," a pug-faced boy said from the middle of the room, "Nakajima stole my pencil today and broke it in half."

A jolt seemed to pass through the class, and several people looked to the back of the room where Nakajima sat with his chin to his chest.

"Is that true, Mr. Nakajima?" Denver asked.

"Yes sir," he grumbled.

"Can you tell us what happened?"

The boy spoke quickly, without looking at the teacher. "Suga pushed me over the chair, so I picked up his pencil off the ground and threw it against the wall. It was no big deal. He started it anyway."

"Liar," someone hissed from the far side.

"Am not," Nakajima shouted back.

"Did anyone else see this incident?" Denver asked, now rising to stand in front of the class with hands drawn across his chest.

"I did," said a girl seated in the back next to the accused. "They were arguing about some dumb comic book. I didn't understand, but I did see Suga-*kun* push him down. And then Nakajima-*kun* broke his pencil." She used the title *kun*, in place of *san*, to show familiarity.

"I see," Denver said. "Mr. Suga, Mr. Nakajima, please come to the front of the class. I don't know what you were fighting about, but I do know that it was not very important and that it was not a very good idea. In my view, both of you were wrong. I want you to apologize to each other right now and then apologize to the rest of the class. This is not appropriate behavior."

There was a pause as the two boys made their way to the front of the room, bowed meekly to each other, and then uttered a barely audible "*Sumimasen*. Excuse me."

"Now everyone listen to me," Denver said, putting his arms around the two boys and returning to his friendly voice. "This is a minor incident, but we want to avoid this sort of thing in the future. It doesn't help our class image."

The students listened quietly.

"This afternoon, as you go through class, I want you to remember our motto, 'Always be courteous. Always be kind.' I don't want to hear about this sort of thing in the future. Now go and enjoy your break."

The students scrambled to their feet and rushed through the open door. Nakajima and Suga hesitated for a moment and then went running after the pack. They had fifteen minutes of recess left before the fifth-period bell.

II

MADE IN JAPAN: NEW YEAR'S EVE AND THE RISING SUN

Praise to Joy, the God descended,
Daughter of Elysium,
Ray of mirth and rapture blended,
Goddess to thy shrine we come.
— *Friedrich von Schiller, "Ode to Joy," 1785*

THE TELEVISION SET was already singing at full volume when I arrived at the Cherry Blossoms' home a little before midnight on New Year's Eve. On the screen, the Tokyo Philharmonic and its background choir were climbing toward the climax of Beethoven's Ninth Symphony. Mrs. C and her two teenage boys were rocking back and forth on the floor, waving their arms and clapping their hands in perfect time to the music.

"Ssshhhh," they whispered when I walked through the door with Mr. C. "This is the best part."

Beethoven's Ninth is to Japan what "Auld Lang Syne" is to the West. This soaring, romantic opus was played in Japan by triumphant American soldiers at the end of the Second World War and has remained ever since as the symbolic music of closings. It is such a perennial best-seller that the Japanese inventors of the compact disc system designed the CD to be seventy-two minutes long so it could hold the symphony in its entirety. In the several weeks leading up to the end of the year, the closing movement of Beethoven's Ninth was performed no fewer than twenty-five times in the Tokyo area alone, and it was played on countless other occasions over the public address system during cleanup hour at Sano Junior High.

At the end of the piece the Cherry Blossom family cheered and the television turned suddenly black. Then, at precisely midnight, a small group of gray-suited businessmen appeared on the screen, standing in the straits between Honshu and Hokkaido — Japan's two largest islands — and proceeded to dedicate the country's newest tunnel. With great earnestness the men raised a toast of sake and drank to the long life and prosperous future of their underwater pass. Next, the cameras cut to the scene of a recently completed bridge between Honshu and Shikoku, hailed as the longest bridge in Japan, where a similar group of men toasted their overwater pass. With this tribute to technology complete, we pulled on our coats, tucked in our scarves, and headed into the woods for our own end-of-the-year celebration.

No holiday captures the tradition, the revelry, and the hybrid spirit of modern Japan better than the five-day midwinter festival of *shōgatsu*, or New Year's. Like Carnival to the Brazilians or Midsummer to the Swedes, *shōgatsu* is a national holiday when workers stay home from work, students go home from school, and city dwellers return to their ancestral homes in the country. For two days before the new year, families scrub their houses and prepare special holiday cards; for two days after, they visit relatives and take gifts to friends and colleagues. But the high point of the celebration, when all the land's afire, begins just after midnight on January 1 with the striking of the New Year's bell.

As one of the few occasions in the year when work stops and families spend time together, the New Year's festival brings out some of the latent religious traditions that still color Japanese life. Our first stop, a fifteen-minute walk from Mr. C's home, would be at a small Shinto neighborhood shrine.

"Have you ever seen a Japanese ghost?" asked Takuya, Mr. C's younger son, an eighth-grade junior high school student. "We might see one tonight. They are *really* scary."

"All the ghosts dress in white," said his older brother, Yuji, waving his arms toward the porcelain moon, "and they float in green smoke. But they don't have legs, so you can outrun them."

"I'm not worried," I assured them as I teetered along on my infirm leg, now finally released from its cast. "They are probably afraid of *gaijin*, just like other Japanese."

Mr. C pointed to an old house set apart in the trees. "My mother was born over there," he said, "and my sister still lives in the house next door. This is my family's land, and we are going to visit my family's shrine."

"Does your family *own* the shrine?" I asked.

"No, but my ancestors have lived in this neighborhood for centuries, and my cousin still serves as the priest. His daughter will probably dance tonight."

Shinto shrines, ranging from sprawling estates in major cities to tiny shelters along rural highways, still proliferate in Japan, reflecting a not-too-distant past when state-sponsored Shinto worship dominated the spiritual tenor of the country. Shinto is a native Japanese cult that honors natural spirits living in trees, rocks, and other objects. It provides no code of ethics or rules about the afterlife but instead stresses a respect for nature and natural phenomena of the earth. Over the years, Shinto evolved to fit its surroundings, developing new rituals for the rice harvest, the fish crop, or other customs important to the towns where it thrived.

About a hundred years ago, when Japan began to reassert its national power, Shinto was heralded as the native theology of the Japanese people. At a time when many people were cutting off their topknots and eating beef in an exaggerated attempt to become Western, Japanese students were learning in school that they had descended from the goddess of the Sun, through the emperor, into an enlightened race. Shinto was used to prove the uniqueness of Japanese culture and justify the country's imperialistic moves into neighboring countries. This recipe of Shinto beliefs mixed with military power formed the ingredients for the Greater East Asian Co-Prosperity Sphere and, later, the Pacific War.

The Allied victory in the war soon ended this scheme and with it the theocratic aspirations of the state. Occupying Americans took control of the government and legally separated church from state. But like many other things the Americans introduced, this separation lasted in name only. A half century after the war, Shinto shrines hold a lingering appeal as the primary havens for neighborhood and family spirits. To this day, from the busiest district of Osaka to the tiniest hamlet in Tochigi, no *shōgatsu* holiday would be complete without a *hatsu-mōde*, a "first visit" to a Shinto shrine.

"There it is," Takuya said, tugging at my sleeve and dragging me down the road. "That's the shrine."

"Ooh, oooooh," Yuji cooed, as if a ghost might descend from the trees.

The vermilion wooden pavilion stood submerged in a grove of pines at the crest of a tiny hill, silent and austere in the faint glow of the moon. The path up to the face of the shrine began beneath the ceremonial gate known as a torii, or bird perch, a post-and-lintel structure with upturned edges.

"Take some small coins," Mr. C whispered, "no more than five or ten yen, and follow me to the entrance. It must be your own money. If I give it to you, it won't work."

We walked under the eaves of the shrine and peered inside at the sacred altar, a scarlet niche decorated with smooth stalks of bamboo, clipped pine branches, and threads of white rice paper.

"Toss the coin into the trough, and pull the bell — hard."

As the coin ricocheted in the wooden hollow, I clanged the copper bell to summon the *kami*, the patron god of the shrine. Following Mr. C, I clapped my hands twice, bowed, and offered a wish for the new year. As we prayed, a white-robed priest shook white paper above our heads, and a teenage girl in a silken white dress danced silently in her bare feet.

This ritual purification — the blond bamboo, virgin white paper, and freshly cut pine boughs — is part of the enduring appeal of this creed. Shinto endures because it celebrates life. Despite a general lack of enthusiasm for formal religion among younger Japanese, most still visit a shrine to mark special, life-cycle occasions — the birth of a baby; the third, fifth, and seventh birthdays of a child; and the ascension of an adolescent into adulthood at the age of twenty. These rites of passage are marked in Japan not by appealing to an abstract, extraterrestrial god but by summoning strength from the spirits of nature, ever present in the trees, the stones, and the land.

"Let's go," Mr. C said, putting his arm around me and leading me away from the altar. "Now, we drink."

The start of the year, like the start of a party, was marked with a toast as well as a prayer.

"This is special *shōgatsu*-sake," he said, pouring a dose of sweet, warm liquor from a wooden bowl into my mouth. "There are gold flakes in this sake. From this night on, you will always have a bit of Japanese gold flowing in your blood."

We left the shrine after several more cups of gold sake and walked over the valley and through the woods back to the Cherry Blossoms' home. Having arrived back at the house, the five of us piled into Mr. C's Nissan Gloria sedan and headed toward the hills west of Sano. After a short drive we approached a gathering of red-and-white-striped tents bathed in a magical yellow glow from a string of paper lanterns. Lazy smoke serpents from vendors' grills wound their way through the light toward the empty black sky above. A festival was under way.

Although it was close to two A.M. when we arrived, the crowd still throbbed with delight. Children pushed their way through the crowd of coats to ogle at candied bananas; a balding man with a two-day growth of beard slapped a slab of squid on a grill and doused it with soy sauce; a grandmother eased her chopsticks into a plate of amber noodles and made a fearsome slurping sound. The glob of noodles disappeared in an instant, leaving two pieces of pink pickled ginger dangling on her upper lip until her tongue came to the rescue. Mr. C bounded from booth to booth, occasionally bursting into song or giggling as he greeted old friends and bowed cheerfully to former students. His sons, with their teenage reputations to uphold, were more somber, but even they drank a cup of warm sweet sake.

We made our way through the crowd and began to climb a narrow stone staircase that spiraled away from the din of the festival to the ponderous seclusion of a decaying temple poised on a cliff above. Having prayed for the future at a Shinto shrine, we would now pay homage to the past at a Buddhist temple.

"Here we must be quiet," Mr. C whispered in my ear. "We do not want to disturb the spirits of the past."

Like many other things Japanese, Buddhism first came to Japan via China. A Chinese priest visiting Japan in the sixth century first explained to Japanese nature worshipers the ideal of reaching a personal salvation in paradise. Buddhism offered individuals a way to escape human suffering by transcending the endless cycle of life and death. By following the teachings of Buddha, any person could achieve a painless state of transcendence, or Nirvana. As Buddhism grew, it split into two divergent streams. An orthodox wing, which stressed the original teachings of Buddha, spread across Southeast Asia, while a new branch, which introduced a pantheon of smaller gods to help individuals reach paradise, spread north through China. This second current, known as Mahayana or "Greater Vehicle" Buddhism, found its way from the Middle Kingdom of China, through the Korean peninsula, to the islands of Japan.

The Japanese never took to the idea that people should adhere strictly to one set of beliefs. Buddhism, like Shinto, proved through the centuries to be quite tolerant and malleable. As Buddhist temples appeared in towns that already had Shinto shrines, and as local farmers who worshiped Shinto gods attended Buddhist funerals, the two beliefs underwent a sort of

Darwinian evolution in which they adapted to fit the needs of their environment. Just as tigers developed stripes to meld better with their surroundings, so Buddhism developed such rituals as prayer before a statue — an act not native to the religion — to suit the traditions of its target audience. Today, the twin ideologies stand side by side in the lives of many Japanese. According to one popular saying, eighty percent of the population say they are Buddhist, eighty percent say they are Shinto, and eighty percent say they believe in no religion at all.

"We have to buy a charm," Mr. C said, pulling my arm and dragging me toward a vendor's window. "This is the Year of the Dragon."

In their potluck creed, the Japanese still abide by the old Chinese calendar cycle, in which every year is named for one of twelve different beasts: snake, horse, sheep, monkey, rooster, dog, boar, rat, ox, tiger, rabbit, and the ultimate lord of the zodiac, the dragon.

"Now," he said, "we pray."

We took our charms and stood before a dark wooden altar, more somber and foreboding than the Shinto shrine had seemed. The columns were thick and gray with age; the wooden floors were splintered with an uneven grain. The air was spiced with burning ash. We clapped our hands twice, as before, and bowed our heads. But instead of throwing money, we lit the wick of a small white candle and placed it on the altar.

"This is for your ancestors," Mrs. Cherry Blossom said. "We want them to know you remember."

"Now," Mr. C said when the flame had risen to its peak, "we have one last thing to do. We must ring the New Year's bell."

For a small fee we entered a small covered arcade and took turns pulling a wooden clapper the size of a baseball bat into the side of a cast-iron bell, sounding a chime that when rung 108 times would clear away the 108 devils that had darkened our souls during the foregoing Year of the Rabbit.

The Year of the Dragon was barely three hours old and already we had pleaded twice at two separate altars, drunk purified liquor laced with gold, toasted the new year and rung out the old, eluded the wrath of Japanese ghosts, and expunged a corps of demons. One would have thought that our chances for success in the coming year were pretty much guaranteed. But Mr. C's boys had another idea.

"Let's go see the first sunrise," Takuya said.

"Ooooh," his brother hummed in eerie, sci-fi tones, "the goddess of the Sun."

"What happens at the first sunrise?" I asked.

"It's the most important time of the night," Mr. C said. "It determines your fate for the year."

"My fate?"

"You'll see." He turned to his wife for approval. "Shall we go?" he asked.

"Yes, let's," she said.

"*Baka yaroo!*" the boys screamed in concert. "This is totally awesome."

An hour later, after a brief stop at home for supplies, our four-door Nissan Gloria pulled into a crowded parking lot at the top of a small mountain overlooking the town. It was a little past four in the morning. Most people below were already sleeping, but in a clearing on this rocky peak, about forty revelers had gathered together to witness the first breath of the Dragon Sun.

The worshipers were huddled around a makeshift campfire, chatting excitedly as they held freshly pared sticks over the fire with what appeared to be marshmallows dripping in the flames.

"What's that?" I asked Takuya.

"That's *mochi*," he said, drawing a similar white block from a plastic bag and plopping it into my hand. "We have some, too. They're like rice balls, but special for New Year's."

He scurried away to pick some sticks as the rest of us took

our places around the circle, where all eyes were trained on the tips of the skewers. Just as one of the white cubes began to dangle from the end of a stick, the roaster would plop the *mochi* into his mouth, howl at the heat, hiss with pain, and finally stretch the blob, taffylike, from his teeth.

"It's hot, it's hot," shouted a woman wrapped in an Indian scarf as she pinched her earlobe with her burning fingers.

"Here, why don't you try some?" a man said, handing me a melted lump.

Moving slowly to avoid the heat, I pulled the puttylike ball to my mouth and sank my teeth into the center. As I chewed, the crowd leaned closer to watch. By this time quite accustomed to such attention, I smiled and waited for their surprised reaction when they discovered that I too could enjoy this treat.

But as soon as the *mochi* reached my tongue, I knew I was in for trouble. This gelatinous paste tasted more like plastic dental x-ray tape than toasted marshmallows. "Be a good Japanese," I thought to myself. "*Gamman* — suffer with dignity."

"Well . . . ," Mr. C said, plopping a block into his mouth and swallowing it as a frog would eat a fly — in one gulp. "How do you like it?"

"It's fine," I mumbled, trying to swallow but feeling the weight of the ball swelling inside my throat. I wondered for a second if this was what a snake felt like after eating an entire rat. But then I reconsidered: a rat seemed oddly appealing.

"This is what Japanese ghosts eat," Takuya chipped in.

"Very funny," I said.

"I'm not kidding," he said. "We put *mochi* in our family shrines at the beginning of every year. Then our ancestors come several weeks later to enjoy our celebration."

"And in Tochigi," his mother added, "we mix *tochi* nuts into our *mochi*."

"It's *tochi-mochi*," Mr. C sang out with a cheery grin.

I smiled through the pain.

Some of the children around the fire had dozed off, and

several of the parents began talking about what they hoped to see later when they went to sleep.

"Remember," said a man with a white cloth wrapped around his head like a warrior and a cigarette dangling from his mouth, "if you dream of a snake, it's bad luck. A fish without scales is worse. But the best dream of all," he said, now turning to address me, "is to catch sight of Mount Fuji."

"*Sooo, desu neee,*" they all chanted, closing their eyes as if trying to call the mountain's face into sight.

"Do you know Mount Fuji?" the man asked me.

"I sure do," I told him.

"Well, if you see Mount Fuji in your first dream of the year, then you enjoy the luck of the goddess of the Sun for the entire year. But don't tell anyone what you see. We Japanese don't tell our dreams."

It was only fitting, I thought, that the goddess Amaterasu should play a leading role in Japanese New Year's lore. The two characters in the name Japan — "Nippon" — literally mean "origin of the sun," and this image is at the heart of the country's interdenominational culture. Although Amaterasu was originally a Shinto goddess, Buddhism responded to her popularity by creating a solar god of its own, Dainichi-nyorai. Over time, these two deities, like their religions themselves, came to be seen as one. Such cross-pollination has led to a set of beliefs — partly inspirational, partly superstitious — that guide the lives of most people.

"Japan is a small country," Mr. C had said to me many times. "Japan is one race," the old man said when explaining the custom of dreaming. "This is the Japanese way," Takuya said about the custom of eating *mochi* on New Year's Eve. According to this gospel, Japan — as a land — is in some ways charmed, in some ways cursed, basically different from others, and ultimately one unto itself. One commentator has called this code "Nipponism." To me, it is simply "Made in Japan."

· · · ·

"The sun! The sun!" someone shouted from the other side of the clearing, and we all rushed to the edge of the hill. As the first ray lipped over the rim of a cloud, the singing and hurrying stopped, and people stood still on the mountaintop to pray.

There was a moment of adoring silence, and then someone behind me shouted, "*Banzai!*"

And everyone answered, "*Banzai!*"

The solitary search for the first dream gave way to the ensemble joy in the first sunrise as cups were raised to the reigning Sun. As I looked out over the sky at this little city tucked away in the mountains of rural Japan, someone tapped me on the shoulder and started to speak in English.

"Excuse me," said the man with the white headband who had spoken earlier of the dreams. "Are you American boy?"

"Yes," I said, "I come from Georgia."

"Me," he said, tapping his finger on his nose like a schoolboy bursting with pride, "made in Sano." He smiled, tipped his cup to mine, and together we turned to face the sky: *Ray of mirth and rapture blended, Goddess to thy shrine we come.*

12

PAUL BUNYAN
AND THE PEACH BOY:
THE JAPANESE COLOR WHEEL

*In the summer of 1945 Japan had two visitors. They
came from the United States of America. One of them
was "Little Boy" and the other was "Fat Man." Who
were they? I think it is very difficult for you to answer
this question.*

— Two Visitors, *ninth-grade English textbook*

THE MORNING BELL sounded at 8:10 A.M. The teachers rose
together from their desks, bowed in unison, and issued a brief
salutation: "*Onegaishimasu.* Please do me the favor of working
hard today."

After nearly six months of making this greeting every morning, I had grown so accustomed to this ritual that I, too, craved
the order — the short, unimposing gesture that united the teachers to one another, reconfirmed their dedication, and established
a tone of courtesy to begin the day. The teachers, like the students, renewed their pledge every day.

Sano was somber in the new year, like an Old West town
boarded up before a fight. The first snow of the season had long
since fallen and been melted away by the sun. The tiny yellow
lights that trimmed the main street before the *shōgatsu* holiday
had been taken down and stored away. At the train station, the
manager had constructed a light blue wooden screen around the
front door to shelter passengers from the wind as they warmed
themselves in front of vending machines that sold hot Georgia

Coffee in a can. Fewer people lingered in the streets in the morning, except for several older men who huddled around the red steam pipe of the sweet potato cart parked at the mouth of the nightclub alley. In the crowded cemetery behind my apartment house, the stone mantles overflowed with small pyramids of mandarin oranges and *mochi* rice balls awaiting the return of ancestral spirits on the fifteenth of the month.

In school, students ran around less between classes, and teachers stayed late almost every day, answering calls from anxious parents who were nervous about the upcoming entrance exams. In Sano, January was a time for crucial studies. In the seventh grade, the time had come for students to learn about the United States.

On the way to class on the second Tuesday of the year, Denver asked if I would give his students a warm-up speech about America. After months of trying to replace fear in the hearts of students with a spirit of adventure, I had mastered this sort of pep talk. Once in front of the class, I drew a map of the United States on the blackboard and began to drill students with the wonders of American Pop Geography: The United States of America is twenty-five times the size of Japan. New York is the largest city in America. Chicago has the tallest building in the world. The president lives in Washington, and Madonna lives in L.A. Chicken is fried in Kentucky, but coffee is not grown in Georgia.

After my speech, we moved on to a review of time, drilling some phrases the students had learned before their New Year's break: "I get up at . . ." "I eat lunch at . . ." "I go to bed at . . ."

Finally we advanced to the text. All three English teachers at Sano Junior High used textbooks from a series written by a Japanese professor of English in Tokyo and approved by the National Ministry of Education. The last chapter of the seventh-grade book introduced America. "The United States of America is a large country between two oceans. There are fifty states in this country. There are many races. Three of them are American Indians, Blacks and Whites. They all speak English."

Denver knew that race was a confusing subject for these students, so he asked me to deliver a short address on the subject. "Use easy English," he warned.

I began with a subject close to home. "In Japanese junior high schools," I said, "there are only Japanese people. Japanese, Japanese, Japanese." I wandered out into the class and plopped my hand on each student's head as if playing a game of Duck, Duck, Goose. "Everybody comes from the *same* country." Back in the front of the classroom, I faced a sea of blank looks.

"Bruce-*san*," Denver whispered in my ear, "the students don't understand that word."

"What word?"

"Same."

"Oh."

I started moving about the class again. "Same, same, same," I sang. "Desk, desk, desk: the same. Hair, hair, hair: the same. Jacket, jacket, jacket: the same."

Slowly recognition began to dawn on the students' faces, and finally a boy called out, "*Onaji.*"

Everyone smiled. The class had learned a new word.

"In Japanese junior high schools," I continued, "all the students come from the *same* country. But in American junior high schools, they come from different countries. We have European people, Indian people, even Japanese people. These people come from different races." As I arrived back at the front of the room, Denver watched with an amused look on his face. "*Sensei* and I are different races. His hair is black; my hair is brown. He is short; I am tall." The students giggled at this observation. "But," I added with great emphasis, "we are both *people*."

For a moment I struggled to think of a way to explain the idea of human equality using the limited one-hundred-word vocabulary of first-year students of English. It came to me in a flash.

"He *gets up* at seven, and I *get up* at seven. He *eats lunch* at twelve; I *eat lunch* at twelve. He *goes to bed* at eleven; I *go to bed* at eleven. We are different races, but we live the *same*."

Anxious to see how the students comprehended this speech, Denver opened the floor to questions. He made a rare exception to accept them in Japanese.

A young girl rose to her feet. "I hear there is racial discrimination in South Africa," she said. "Is this a problem in America?"

I was stunned. After all, she was only a junior high school student . . .

"Yes," I said, "we have this problem in America, but it is not as bad in the USA as it is in South Africa." Again looking for a way to illustrate this problem, I caught sight of the blackboard.

"When my father was a junior high school student in Georgia, there were only white people in his class." I drew two schoolhouses on top of Georgia: one I colored white; the other I left dark. "But when I was a junior high school student, there were many kinds of people in my class. Black people, white people, Hispanic people, even Japanese people." I drew two arrows and joined the schools as one. "Now we study together."

Next, a boy stood up and introduced himself. "Hello. My name is Hoshino, but my nickname is Yotch. I hear that discrimination is a problem in Alaska. Is that true?"

One of the most challenging aspects of teaching in Japan was the questions I was called on to answer as a representative of my country, and, indeed, of the entire English-speaking world. These queries ranged from the linguistic ("Why is someone from the Netherlands called Dutch?") to the curious ("What have you done for world peace lately?") to the culinary ("I like hotcakes very much, but they are not very tasty. What advice can you give me to make hotcakes more delicious?"). The civil rights ambassador in one class became the itinerant gourmet in the next.

Here I was in a seventh-grade classroom in rural Japan

being asked by a twelve-year-old boy named Yotch to comment on the racial discrimination faced by Eskimos in Alaska. Although the thought crossed my mind, I could not tell him that it was as far from my home to Alaska as it was from his home to Tibet. I could not tell him that the problem was under advisement by civil rights commissions, political panels, and government agencies. Instead, I had to answer the question directly — and in "easy English" to boot — because this was the reason I had come to Japan.

"Eskimos live in Alaska," I said. "They have lived there for many years. Many of them speak a different language and have different customs from others who live nearby. When people do not talk to one another, sometimes they are afraid to trust. We must learn to talk with people who are different from us."

As I finished my sermon, which most of the students clearly did not understand, I walked over to Yotch to thank him for his question. I stuck out my hand; he grabbed it and squeezed hard. Finally dropping it from his grasp, he turned to the other students and raised his arms in the air like a prizefighter.

"*Kakko ii*!" he shouted. "Pretty cool!"

Everyone understood this.

Sooner or later, almost every conversation I had on the differences between Japan and the rest of the world returned to a singular issue: race. Seemingly unrelated facets of life in Japan were explained as being the result of the country's exclusive stock. Why do Japanese students wear uniforms? "Because Japan is one race," a teacher told me. Why does Japan have no drug problem? "Because we are one people," a doctor declared. While homogeneity has been a major factor in Japan's development, surely it cannot explain all the eating, drinking, sleeping, shopping, schooling, mating, and dressing patterns of the country. This catchall has clearly caught too much.

The notion of racial distinction has long been central to

education in Japan. For as long as Japanese parents have been teaching their children about the rest of the world, they have been drawing comparisons between themselves and foreigners based on size, strength, and color. Early visitors to the country were labeled "red hairy barbarians." Later foreigners were described as "smelling like butter" because of their association with beef and thus with cows. During the Second World War, while Americans dubbed the Japanese as yellow apes and buck-toothed beasts, the Japanese tagged Americans as evil demons and horse-tailed brutes. "Every war bond kills a Jap," the Americans hailed; "Every savings account kills an Allied devil," the Japanese rejoined.

If forging racist slurs in wartime was simple, eliminating them in peacetime has proved more difficult. The United States and Japan stand as allies today approaching half a century of cooperation. Yet they are also competitors, and in times of increased tension the racist images of wartime are quick to resurface. Given this burden of history, teaching about race has become one of the fundamental challenges facing the Japanese school system. How they treat this issue, above all others, will determine the success of their efforts to internationalize. For teachers, this question shadowed all of their classes: how could they balance students' age-old inclination to love their country with the modern imperative to live and work in harmony with the black devils, vile demons, and other *gaijin* across the sea?

"Mr. Bruce," Mrs. Negishi said to me as we were walking toward class that afternoon, "this year is almost over. Soon my students will take their entrance exams and move on to high school. This is their last chance to hear a handsome native English speaker like yourself. Today, they want you to tell them a story."

"What kind of story?" I asked.

"A folk tale."

"Any particular one?"

"Yes," she said, "Paul Bunyan. That's their favorite one."

The textbook series that teachers used in Sano made considerable effort to give students a well-rounded view of life in the English-speaking world. After the seventh-year book introduced the United States, subsequent books contained sections on hand gestures in America, seasons in Australia, and weather in Britain. For students, the most popular chapters described folk heroes in these countries. In January, at the same time the seventh graders took up race in America, the ninth graders began to read about that lovable, lumbering American folk hero, Paul Bunyan.

I began my talk, as always, by sketching a cartoon map of the United States on the blackboard: five lakes drooping down from the north like an udder, a horn for Maine, an oversized lobe for Florida, and a sweep of the chalk along the bottom for Texas and California.

"Mr. Tanaka," I called, "what is this?"

By now the students were becoming accustomed to these random outbursts of interrogation from the teacher. Early on I realized that my teaching style would be different from that of most of my Japanese colleagues — there would be more questions, more gesturing, more pressure on the students to participate in class. Although many of the students were originally shocked, in time they warmed to this "American" style.

"America," several students called out, without waiting to be identified.

"Correct. *The* United States of America."

"*Thheeee* United States of America," they said, mimicking my enunciation.

I picked up some colored chalk from the blackboard trough and began shading different regions. "This is the East of the United States of America. New York is in the East." I shaded this part yellow. "This is the West. Los Angeles is in the West." I colored this part blue. "In between the East and the West is the area we call the Midwest."

I filled in this zone with green and explained how one hundred years ago many men moved from the East to the Midwest to cut down trees to build houses, boats, and even students' desks. To illustrate this new phrase — "cut down" — I grabbed a tennis racket from a student's locker in the back of the room and pulled a student to the front of the class.

"He is a tree. I am a woodcutter. Now I am cutting down wood." Within seconds the student had been felled to the floor and everyone had acquired a new phrase.

"Many of the woodcutters in the Midwest lived in one big house," I explained, "and every night they sat around a big table, drank beer, and told stories.

" 'Today I cut down ten trees,' the first woodcutter would say.

" 'I cut down *fifty* trees,' another would say.

" 'I cut down *one hundred* trees,' a third man would add.

"Finally, to stop this game, they began to tell stories of the biggest and strongest woodcutter of all, who could cut down *one thousand* trees in a day. His name was Paul Bunyan."

The students cheered when they heard his name. "Cromartie, Cromartie," two boys screamed, calling the name of a popular American baseball player in Japan who slugged home runs for the Tokyo Giants.

When I had finished, Mrs. Negishi asked the class if there were any questions.

At first no one spoke, but several students peered toward the back of the room, where a junior high school setup was clearly under way. Slowly a boy stood up with a vivid blush and asked, "Did Paul Bunyan have a girlfriend?"

"I don't know," I answered. "But he did have a giant blue ox named Babe." Mrs. Negishi laughed, but the students didn't, and my joke got lost in a scramble to ascertain the Japanese word for ox.

"We want to thank Mr. Bruce for explaining the story of the

American folk hero Paul Bunyan," Mrs. Negishi said, resuming her place at the head of the class. "Now it is our turn to explain to him the story of a Japanese folk hero. Is everybody ready?"

Several girls jumped from their seats and scurried to the back of the class. Some boys began rummaging through knapsacks, pulling out pieces of paper and assorted props. Within seconds an entire cast of ninth-grade thespians appeared before the class, dressed in monkey masks and devil's horns and bearing a collection of cartoon drawings.

"Attention," one of the girls said to quiet the group. "Now we will tell you the story of Japanese folk hero Momotarō, the Peach Boy.

"Once upon a time there were an old man and an old woman who lived in Japan." From stage left a girl held up a pastel sketch of an old man and an old woman; from the right, two boys dressed in bathrobes and slippers stepped forward. One wore a cotton-ball beard and carried a cane, while the other showed off a lipstick smile. The class applauded.

"One day the old lady was washing clothes in the river when she discovered a peach in the water. She picked up the fruit and took it home. Later, when she cut it open, she found a little boy inside. She named him the Peach Boy."

A small boy emerged in full peach panoply from behind the wall of actors and jumped into the waiting arms of the old woman. The audience squealed in delight. These boys had already mastered the exaggerated art of Japanese camp and would, no doubt, make fine *kara-oke* artists someday.

"When the Peach Boy grew older, he became very strong. He decided to save his country from the evil devils across the sea who invaded Japan every year. Before his trip, the old lady gave the Peach Boy a bag of the most delicious dumplings in all of the land. Then he went away."

"On the road the Peach Boy met a dog. 'What do you have in your bag?' the dog asked. 'Dumplings,' the Peach Boy said.

When the dog ate one of the dumplings, he thought it was very delicious, so he went with the Peach Boy to kill the devils."

This part of the play was acted out in front of the class, complete with actual dumplings from a plastic shopping bag. After winning over the dog, the Peach Boy recruited a monkey and a pheasant in the same way.

"The Peach Boy and his three friends then traveled together across the sea to kill the evil devils. When they arrived at the island, the dog screamed, the monkey jumped, and the pheasant flew into the air. Together with Peach Boy they killed all the foreign devils."

The actors went racing through the rows of desks during this last part, hitting students on the head and screaming attack slogans in Japanese. By the end of the run, the students were in stitches, the teacher from the neighboring class had begun banging on the wall for silence, and the Peach Boy was proclaimed a hero. "The Peach Boy returned to Japan with a bag of treasures he had taken from the devils, and all the people gave him a parade. This is the end."

The actors bowed together as the final bell rang. Mrs. Negishi thanked the participants and brought the class to a close just as a swarm of neighboring students poured into the room. With all the commotion I had no chance to ask my question, "Did the Peach Boy have a girlfriend?"

Not long after I heard this tale at school, I learned of the peculiar role this story played in recent Japanese history. During the Pacific War, the tale of Momotarō became a key plank in the government's propaganda strategy. With its heroic native son fending off fierce foreign devils "across the sea," the story was the ideal paradigm for the Japanese war with the Americans. During the war, the story was included in the elementary school language textbooks, and Momotarō appeared in a host of wartime publications, including cartoons, magazines, and animated films.

The diminutive Peach Boy also starred in a war film entitled *Momotarō and the Sky*, in which the boy-hero sacrificed himself to save a flock of imperiled penguins at the South Pole from invading eagles. In one film Momotarō was transfigured into a stern military commander demanding the surrender of British troops at Singapore. His retainers — the dog, monkey, and pheasant — were loyal military officers promoting Japan's Asian Co-Prosperity scheme.

"The Japanese think they are small and vulnerable," Denver said to me one night after I learned about Momotarō. "This is the Japanese sense of history. We call it *higaisha* — one who receives pressure." He ground his thumb into his hand to illustrate the feeling. "The Japanese feel as if they are always under pressure from the rest of the world. The opposite is *kangaisha* — one who applies pressure. That is the United States."

For much of its history, Japan has felt victimized by external forces. In the early seventeenth century the ruling Tokugawa clan banned Christianity from Japan, believing that it was corrupting the population. In 1636 the regime went so far as to prevent Japanese who had lived abroad from returning home, for fear that they had been tainted by foreign influence. For the next two hundred years Japan closed itself off almost entirely from the outside world. Ports were sealed; travel was banned. But with international trade growing more important, this isolation could not last forever. In 1853, in one of the most mythologized events in Japanese history, the American commodore Matthew Calbraith Perry sailed a small flotilla of ships into Uraga Bay outside the capital and demanded that the shogun open Japan's markets to foreign countries. This expedition, nicknamed the "black ships invasion," became a symbol of foreigners forcing their will on the hapless Japanese. Today, more than a century later, to raise the specter of "black ships" is to sound the *higaisha* call to arms.

While politicians can use this call to arouse the people, teachers are in a much more sensitive position. Ideally, schools should present a more balanced and less inflammatory view of history. But in recent years an international controversy has sprung up over how Japanese textbooks depict the behavior of the Japanese during the Second World War. The Japanese have tried to present themselves as victims of the West. They point out that the Greater East Asian Co-Prosperity Sphere and its promise to "throw off the yoke of Imperialism" appealed to the peoples of Asia who had suffered under European colonial rule. Governments in South Korea and China have complained about such accounts; they claim that the oppression felt by Asians who lived under brutal Japanese domination has been left out.

The history textbook used at Sano Junior High devoted five pages to the war in the Pacific. One page discussed the early years, including the attack on Pearl Harbor; one page talked about the middle years and the fighting on the Pacific islands; three pages reviewed the American fire-bombing of Tokyo and the dropping of the atomic bombs on Hiroshima and Nagasaki. By contrast, the account of the war dismissed the Japanese invasion of Nanking, including the alleged massacre of two hundred thousand people, in less than three lines. The Japanese invasions of Burma, Malaysia, and Indonesia were not mentioned at all.

"Every Japanese junior high school student knows about Hiroshima and Nagasaki," Denver lamented, "but they don't know about Nanking."

"Then how do you know it?" I asked.

"I learned it at my university, but not everyone has this chance."

Not only did the history textbook dwell on the lessons of Hiroshima, but the English textbook took up the issue as well. The penultimate chapter of the ninth-grade book, the one directly preceding the account of Paul Bunyan, explored the subject of nuclear holocaust. The title of the chapter, "Two Visitors," referred to the two bombs the United States dropped on Japan in the early days of August 1945.

"Little Boy" came to Hiroshima on August 6th. "Fat Man" came to Nagasaki three days later. When "Little Boy" fell from the sky above Hiroshima, he blew up and gave out a strong light and great heat. The heat was so strong that it reached 7,000 degrees Centigrade. Thousands of people were killed. Can you guess how many? More than two hundred thousand.

The story went on to discuss the danger of assigning pleasant names to harmful things.

Some people called the bombs "Little Boy" and "Fat Man." They are charming names, aren't they? They were the names for the atomic bombs dropped on Hiroshima and Nagasaki. A name is a name. It is not a thing. I think it is important to remember this. A thing with a charming name can sometimes do a cruel thing, like "Little Boy" or "Fat Man."

The purpose of the chapter was not to discuss history but to teach the simple phrase "I think it is . . . ," as in "I think it is important to remember this." But by choosing this story, the author seemed to be reinforcing the popular notion of Japan as victim.

"The Japanese have a collective historical amnesia," said a friend who teaches history at the University of Tokyo. He blamed the secondary school system. "Most students are not taught the history of Japan's invasion into China and other nations. They are taught about the poor times after the war. They are taught that their parents worked hard to overcome these difficulties. We are fearful these difficult days will come back."

This background makes the Japanese move toward internationalization, or *kokusaika*, all the more difficult to attain. If *kokusaika* is to be achieved, students must learn to look beyond their fear of foreigners. They must make special efforts to embrace new cultures — American, European, African, and

Asian — rather than retreat into the protective shell of their own tradition. Unfortunately, at the time when schools should be refocusing their gaze outward, the government seems to be encouraging students to train their eyes inward on their native past. Even as schools teach English to their students, politicians go on national television warning everyone not to grow "lazy" and "sloppy" like American workers. Even as schools tout "international" thinking, they insist that all students learn martial arts, the haven of the "Japanese spirit." One school near Sano was part of a nationwide program that required every child to study *kendō*, a traditional form of fencing with rigid mental training and discipline.

This legacy of parochialism is nowhere more evident than in the problem of "returnee" children, students who return to Japan after living abroad for a time and attending foreign schools. These students, with their superior foreign language skills and firsthand knowledge of the outside world, might be expected to be in the vanguard of the *kokusaika* movement. Instead, many are ostracized because of their limited knowledge of Japanese and their perceived "foreign" tastes.

At the beginning of the second term in Sano, a girl who had lived for seven years in Peru returned to school to take her place in the ninth grade. Mrs. Negishi was concerned about how this girl was adapting to her new surroundings and asked if I would speak to her in Spanish, which I had studied briefly in college. Many of the students in school, Mrs. Negishi said, were treating the girl as an alien because her Japanese lagged behind the others' and her accent sounded strange. The following week, without warning, Mrs. Negishi called the girl to the front of the class one day.

"Mr. Bruce will now speak Spanish with you," she announced to the girl.

The entire class turned to watch the girl as she made her way to the front of the room. I felt uncomfortable that this spotlight was being thrust on her, but I managed to offer a short greeting. The girl balked at the sight of my outstretched hand and stopped short at the first row of desks. Her fingers whittled at the hem of her jacket. Her head dropped to her chest.

"*Gamman*," the teacher encouraged, but the girl refused to move. Tears rolled down her cheeks. The pressure of being alone proved to be too great for her to accept. After standing in the center of the room for several seconds, she returned to her seat in silence.

This episode struck me as symbolic of the difficulty of blending cultures in Japan. Obviously few of the students in that room had adjusted well to the new girl in their class. Few had accepted this outsider in their *kumi*, even though she looked, dressed, and even acted the same. "The Japanese are good at absorbing foreign culture," a leading newspaper editorialized soon after this, "but decisively lack the capability to coexist with it." The students I taught had yet to learn that it takes all colors on the color wheel to make a perfect circle. What they needed most was a new folk hero to lead the way of their growing international movement, a figure less threatening than a foreign giant and more open-minded than a provincial child: in short, a cross between Paul Bunyan and the Peach Boy.

13

KEEPING THE FIRE ALIVE: TWIN WINTER ESCAPADES

Jog on, jog on, the footpath way,
And merrily hent the stile-a:
A merry heart goes all the day,
Your sad tires in a mile-a.
— *Shakespeare*, The Winter's Tale

THE FIRST TIME I PLANNED to visit Tokyo, back when the autumn leaves were still aflutter and the wind was sweeping through Sano, I asked Mr. C how to get there.

"It's too *far*," he said, wrinkling his nose, squinting his eyes, and twisting his face in concern. "You must change trains. I can't go with you. I'm not sure you can do it alone."

Next I approached Arai-*san*.

"Tokyo?" she gasped, staring down at my leg, which was still wrapped in its cast. "You'll get mugged. They'll hit your leg. In the subway they're really mean."

Finally I gave up and called Cho.

"It's easy," he said. "I'll draw you a map. I'll ask my friend Hara to meet you."

Tokyo is a fountain of youth — a rich, commercial paradise brimming with juvenalia and overflowing with tides of voracious young adults with a seemingly insatiable appetite for consumption and mirth. For older people, like Mr. Cherry Blossom, Tokyo is a modern-day den of iniquity, but for younger types, like Cho and his college pal Hara, the city is a mecca of fantasy, offering escape from social pressure within its neon lairs of pleasure tucked away in the maze of skyscraping towers.

Hara, who was born in Sapporo, on the northern island of Hokkaido, first came to Tokyo for college, where he joined the same club as Cho. After graduating, he spent two years working for the Japanese Foreign Ministry in Cairo and then returned to Tokyo — literally, the "Eastern Capital" — to work for a major industrial bank. He was younger than Cho, and taller, with blow-dried hair, a clipped mustache, and a sizable beer belly. He had a fondness for seersucker suits.

"Ah yes . . . Georgia," he said to me over dinner one night in an English rich with the lilt of a British country baron. " 'Georgia on My Mind' . . . *Gone With the Wind*. Yes, I like the South. Did you know my bank is thinking about buying some property down there? Of course we would only buy the best hotels; that way we know the return would be high."

"Why the South?" I asked.

"Real estate is so expensive in Tokyo, so we have to look elsewhere now. In the South, the food and the culture are more like they are in Japan. The people are also more polite."

He poured beer into my glass; I did the same for him, and he downed his draft in one gulp.

"By the way," he said with chopsticks poised to take another bite of fish, "do you like Japanese girls?"

"Are they different from American girls?" I said.

"Of course," he said. "They're better. They are pretty to watch, and they don't talk back. The next time you come to Tokyo I'll find you a Japanese girl. I'll take you on a *go-con*."

According to custom, there are three ways for boys to meet girls in Japan. The most traditional method is to have an arranged meeting, or *omiai*, in which both sides investigate each other before sitting down to tea. The most romantic method is to fall in love, a route that is fashionable among younger people but still considered flighty by conservative types. The most renegade method is for boys to hang out in bars and pick up girls, in a popular pastime known as *nanpa*. These three techniques, however, proved not to be enough for the thousands of young people like Hara who first moved to Tokyo for college and then decided to stay for work. For these people a new form was needed: one that would unite people accustomed to proper introductions, get them out of their one-room apartments, and exploit the advantages of Tokyo's thriving nightclub scene.

What they created is called a *go-con*. *Go* is an abbreviation for the Japanese word for "meeting," and *con* is short for the English word "companion." The *go-con*, like its name, is a cross-cultural hybrid that resembles a group blind date. The rules are quite simple: a boy and a girl who are friends, but not lovers, agree to host a soirée. The boy brings two boy friends; the girl brings two girl friends; and the six — or eight, or twelve — go out to paint the town red and pair themselves off. The *go-con* preserves the security of the arranged meeting while introducing the added element of choice, which has seeped into youth culture from the West. In mid-January, on the day after the national holiday when ancestors return from the afterlife to claim their New Year's *mochi* treat, I got a chance to venture down this modern "Way of Meeting" with Hara and some of his friends.

I arrived at the appointed street corner in the ritzy Roppongi district of Tokyo a little after eight o'clock on Friday night. In addition to me, Hara had invited his friend Azuma, who worked as a cameraman for one of Japan's major television networks, a highly respected and well-paying job that boasts a higher than average glamour quotient. Accordingly, Azuma came dressed for our *go-con* in tapered leather pants and a tight, zebra-striped sweater. His hair was greased back against his head and a small, provocative curl hung down over his forehead, sort of like Michael Jackson, but macho. Call him Prince Charming.

Hara and Azuma seemed ideal for the *go-con* game, except,

as I soon learned, neither one was playing. Azuma already had a girlfriend, who would be our hostess tonight, and Hara was engaged. The night was only ten minutes old, and we were off to an unpromising start.

We made our way to the first site, a second-story restaurant with dark benches and thick tables, and sat down to wait for the girls. (In using this term I defer to the linguistic customs of the country. In Japan, a girl is a "girl" until she gets married, at which point she becomes a "mother.") After almost an hour the girls' team finally arrived from work. The first to sit down on the bench was a slim, regal goddess with black satin hair that tickled her waist, a black silk dress, and deep cinnamon eyes. Her name was Reiko, but call her Sleeping Beauty. She tucked her purple shoes under the bench and began chatting across the table with Prince Charming. I assumed she was his girlfriend and felt a tinge of regret.

Next, an icy woman sat down with scarcely a hello, drew a cigarette the size of a chopstick from her bag, and began punctuating her comments with a cocked, smoking wrist. Kana was her name. When she mentioned that she worked for the Tokyo office of a large Wall Street bank, I asked her what she thought of her American colleagues. She turned toward me and said, "I don't think I am going to tell *you* what I think of *them* until I decide what I think of *you* first." Call her the Wicked Stepmother.

The last girl to sit down at the table was the one person whose name I never learned. In fact, in the course of the entire evening, the only thing she said to me was, "Wow, you've lived in Japan only six months. You speak Japanese so well. I lived in America for two years, and I can't speak a word of English. Gee." Call her Dopey.

Once all the players had gathered around the table, our troubles began. I was speaking with Azuma across the table when suddenly he turned red and shouted for everyone to stop. Turning to Sleeping Beauty beside me, he said, "Reiko, you talk to Bruce. Kana, you sit next to me. And Hara, you talk to her," the unnamed woman at the end. As soon as this shuffle was complete, Azuma began screaming at Kana. "Why were you late?" he demanded. "We sat here for forty-five minutes waiting for you to come. Who do you think you are?" Unconcerned, she stared quietly into the room and blew smoke into his face.

Meanwhile, I turned to talk to Reiko, now encouraged by the revelation that she was not the object of Prince Charming's affection.

"Is this a fight?" I asked her.

"Wait a minute," she said, "I'll find out." She leaned across the table and tapped Kana on the arm. "Excuse me, is this a fight?"

"Yes," said the Wicked Stepmother.

"How long will it last?" asked Sleeping Beauty.

"About twenty minutes," said Prince Charming.

"Okay."

Sure enough, they continued this heated discussion for the next twenty minutes, while the rest of us drank beer, sampled salted shrimp and spiced cabbage, and pretended not to notice them. When the appointed time arrived, I looked at Reiko, she turned to Hara, and he spun toward Azuma and asked, "Are you finished?"

"For now," he replied. "We'll continue later."

And with that, we all returned to our original seats.

Like its cousin the *enkai*, the *go-con* has an established agenda. First, the guests meet at a bar for food and drinks, then they move to a disco, and finally they stop off at a late-night coffee shop before breaking up for the night. In addition to having a poorly matched clientele, our *go-con* suffered from an unfortunate choice of locales. At our first place we received unusually brusque service. The waiter had told us when we arrived that we would have to leave within two hours. After the forty-five-minute delay and the twenty-minute fight, we barely had time to slip down some raw fish before we were forced to move out. Our second stop, an underground disco called Lexington Queen, was equally uncordial and even less conducive to chitchat.

The inside of this subterranean pleasure palace was filled with rotating mirrors, fish-eyed video screens, and computerized lights that descended from the ceiling, blinked nervously, and spun around in circles. For all the hydraulic histrionics, however, the room had no empty tables for sitting and no vacant corners for standing. We could not even lean against the green velvet walls because they were covered with photographs of all the famous people who had once visited this place: Huey Lewis, Billy Idol, and a local group, the Twin Devils. With no place to linger, we sought the dance floor.

If the first stop of a *go-con* is for getting acquainted, the second stop is for staking claims. In normal *go-cons*, Hara explained, the men huddle early in the evening to divvy up the girls. At our *go-con*, however, no such coordination occurred. Left to my own devices, I decided to make a small advance toward Sleeping Beauty. Unlike in erstwhile days, when the subtle suitor could send his sweetheart a letter — "It's winter; the nights are cool; my body warms for you" — the savvy suitor of today must make his intentions known through suggestive and alluring gyrations on the public stage. With this in mind, I tried to position myself directly across our tight dancing ensemble from Sleeping Beauty to maximize visual contact. Unfortunately, just as I had maneuvered myself into the perfect flirting position, two young teenage nymphs came writhing up to me, practically bursting out of shrink-wrapped dresses that clung to their bodies like milk to a spoon. They winked at me, dropped their eyes in contrived shyness, and intercepted my provocative gestures, mistaking these advances as directed toward them.

Within moments our *go-con* circle had begun to disintegrate. Azuma, by now fighting again with Kana, disappeared into the swarm of people and began dancing with other girls. Hara, somewhat removed from the various intrigues, disappeared into the men's room to telephone his girlfriend. This left me alone with our troubled trio of bachelor girls — one thwarted lover, one silent beauty, and one nameless girl who danced by herself. Our *go-con* had gone awry.

After two hours of crisscrossing the disco like characters in a Chekhov play — "Have you seen Kana?" "Is Reiko still here?" "What was that girl's name?" — we finally managed to assemble at the door and make our way outside. The marquee above our heads said, "WELCOME TO THE LEXINGTON QUEEN: GREAT MYSTERY IN THE EFFEMINATE WORLD."

Our final stop was a tiny Art Deco coffee shop with plastic lilies on the table and gray vinyl tile on the walls. For a moment we seemed poised to forge order out of our chaos. Azuma sat next to Kana; Hara sat next to the unnamed woman; and Reiko sat next to me. Unfortunately for all, the conversation took a turn to the West.

"Now, Kana, you have to tell us what you think of Americans," Hara said after we had placed our order.

"Do you really want to know?" she asked, glancing in my direction.

"I think I can handle it," I replied.

"Okay," she said, "you asked for it." She pulled another cigarette from her bag, and Azuma lunged to light it. "I think they are lazy. These smart young men come here from Wall Street, and all they do is complain about us. They have no manners, and they show no respect. They don't even try to understand Japan. All they talk about is sex and money."

"What don't they understand?" I asked.

She took a drag and tapped the ashes in a tray. "They don't understand that while they are out playing, we are all working. They go out to nightclubs around Tokyo, and they think that's the real Japan. It isn't. Do you know why I was late tonight?

Because my boss — my Japanese boss — said I had to work. The Americans had all gone home."

"But you're *always* late," Azuma said.

"Be quiet," she snapped back. "At least I'm not rude."

"Why are you so angry?" Reiko asked, coming to my defense.

"Japan used to be a poor country," Kana continued. "After the war, we were the slaves of America. But we're not poor anymore, and we don't have to be slaves of anyone. Now we are rich, and soon it may be your turn to be working for us."

"Okay, enough," Azuma interrupted. "Don't listen to her. She's just drunk. She doesn't know when to keep her mouth shut."

"He asked me what I think," she said. "All I did was tell him the truth."

The "truth," in this case, landed like a wet towel on our love circle, but I was happy not to get submerged in a debate on the relative virtues of Americans abroad. Ten minutes later, long enough for me to spill my glass of raspberry liqueur all over the oyster pizza, we decided to end our *go-con*. We left the café and headed toward a deserted side street in search of taxi cabs. Hara and Azuma immediately walked ahead, leaving Kana behind with her friends. I walked in the middle, alone, until Reiko appeared at my side.

"Why is it that Japanese men always walk ahead?" she asked.

Should I tell her that most American men aren't like that? I thought to myself. Should I tell her that I'm not like that? Should I tell her that I'm one of those "Sensitive Men" who care only for the feelings of others? Or should I defend myself as a man?

"I don't know," I said meekly. "Japanese custom?"

"I think it's stubbornness," she said.

Then suddenly, out of nowhere, a jolt of boldness struck me. "By the way," I said, "do you agree with Kana?"

"About what?"

"American men."

"Some of them. Most of the Americans we work with don't care about Japan. They don't want to understand our hearts. But I think there are many Americans who do care about how we feel. I hope you are one of them . . ." For the first time all night, a spring of hope.

At the corner we faced the final task of dividing ourselves into couples for the ride home. To my surprise, Hara took up my cause with Reiko, encouraging her to take me home. "Your apartment is much bigger than mine," he said. "And he is so tall. Why don't you invite him home?"

She didn't answer, content to let the request stand for a moment more.

The first cab came. Dopey got inside but left the door ajar. With great dramatic flair, the Wicked Stepmother marched to the open door and waited for a sign from the Prince.

A second cab came. Sleeping Beauty got inside, but again the car did not drive away.

A third cab came, and Hara stepped toward it.

For a moment, three of us stood waiting at this fork in the woods — Prince Charming, the Wicked Stepmother, and the tall, dark stranger from across the sea. The first to act was Hara, who called from the last taxi cab, "Bruce-*san*, you're coming with me." I moved toward his car, and Sleeping Beauty drove off into the sunrise without even waving good-bye or telling me her telephone number. Meanwhile, Prince Charming and the Wicked Stepmother were still locked in their trance.

"Go ahead," Hara called to his friend, "say something. At least tell her good night."

After another tantalizing moment, Azuma stepped forward, jumped into *our* cab, and without so much as a glance over his shoulder ordered the driver to take us away.

"Oh no," Hara cried. "What a catastrophe."

· · ·

Back at his studio apartment, Hara explained the history behind our failed evening. He told me about successful *go-cons* in the past, when he had met girls who had become passing flings, live-in lovers, and even his fiancée. He told me about Azuma and Kana and their stormy past, with an assorted collection of dramatic fights in public places, followed by romantic make-ups in distant hideaways. He told me about the new era of love in Japan, in which sex comes first and marriage later. "You meet, you talk, you have sex — just like in America."

But then he told me he was tired of this life.

"When I was in university, we had parties like this all the time," he said. "I got money from my parents once a month, and I spent it all in one week. I lived on beer and tuna fish. Even today, the only glasses I have in my house came from *kara-oke mamas.*"

He opened his refrigerator and offered me a drink. The only items on the shelves were a half-empty bottle of brown Chinese tea and two cans of NCAA Isotonic drink.

"I want to have a home, a wife, and a family," he said, lying back on his unmade bed. "I think I'm outgrowing the city. You see that picture over there?" He pointed to a small framed photo of a group of men on a ski slope. "That's where I'd rather be. I'm ready to go back to the country."

"Irashaimase, irashaimase," the old woman said as she sank to her knees, grazing her fingers on the mat at our feet and dipping her head to the floor. "Welcome, welcome. It is late, you must be weary from the drive."

In early February, two weeks after my thwarted *go-con*, I went back to the country with Hara, Cho, and some of their friends for a reunion at the spot they had been visiting since they first went away to school. Cho, Chieko, and I had left Sano early Friday evening and driven two hours west from Tochigi through a mountain pass toward the Sea of Japan. We arrived after midnight in a small town in Niigata Prefecture at an inn named after its homeland, Yukiguni, the "Snow Country." The Yukiguni is spiritually about as far from Tokyo as one can go and still be in Japan. An isolated pocket of mountain villages on the western side of Honshu island, the Yukiguni is to winter what Nikko is to fall. In the peak of the season, snow settles several feet thick over the region, piling in milky dunes atop black tile roofs and gathering in mounds along every thoroughfare. Instead of the glass forests of Tokyo, here open fields outnumber the homes and only evergreens scrape the sky. Long ago, wealthy lords from the capital brought their mistresses here for furtive weekend trysts; now college boys bring their girlfriends for several days on the slopes.

"Please come inside," the woman said. "Get warm around the fire and have some tea. The others have been waiting."

We stepped out of our shoes and through a sliding door into a large, matted room, darkened by worn wooden panels and heavy, knotted beams. Faded prints of pine trees hung from the walls, stained by spots of dripping water which had paused on their way to the floor. The cold air from the room stung my face. I looked but could see no fire.

"You're here," Hara called, dragging his legs from beneath a quilt-covered table. He bowed to Cho, his *sempai*, or senior, at college, greeted Chieko, then slapped me on the back. "Bruce-*san*," he said. "It's been a long time. I want you to meet my fiancée."

Emiko Kawakami was small and perky, with long hair and a marshmallow nose. With her shocking-pink sweater and lipstick to match, she looked like the flouncy models on Japanese television who pitch everything from mouthwash to life insurance with little more than a wink and a bow.

"Nice to meet you, Mr. Bruce," she said with a slight giggle. "Are you related to Bruce Lee?"

"Um, he's my father," I said.

She thought for a moment, then turned toward Hara. "*Uso*," she said. "He's kidding me, right?"

From behind, another man stepped forward.

"Good evening," he said. "My name is Komaba. This is my younger brother, Tomo."

Komaba, who had also been a member of the international club, worked alongside Hara in the same Tokyo bank. He was clean-cut, with a button-down blue Oxford shirt and a bright green down ski vest. His younger brother, although close to twenty, still wore his hair in a junior-high-school-style crew cut. After the usual bows and how-have-you-beens, we lapsed out of formality and sought refuge under the quilt.

I pushed my legs underneath the table, and my feet fell into a hole in the floor and came to rest on what felt like a fire. I stole a glance under the quilt and saw that the heat came instead from a bright red bulb. This was an old-fashioned *kotatsu*, Cho explained, a Japanese hot seat. The earliest *kotatsu* consisted of a cavity carved into the foundation of a house in which a charcoal fire could be embedded; modern technology has replaced the live fire with an electric heat lamp. Since most Japanese homes have no central heating, these toaster tables serve as the main source of heat for many people. Even families who have kerosene heat still have a *kotatsu* for central seating.

After a moment the old woman appeared from behind a curtain and set a small cup of green tea in front of each person. Her face, like the walls, had wrinkled with age, but her smile softened the lines. She settled on bended knees just off to the side of the table and began asking questions of her guests, like a seasoned *kara-oke mama*. A year had passed since their last visit, she said. Had anyone taken a trip? Was anyone planning to get married?

"I went to Australia on business last summer," Komaba said, "and I'm moving to our London office in April. I'm afraid this will be my last year for a while."

"London!" the woman gasped. "Sounds dangerous."

"Not really," he said. "It's not like New York. I won't need to buy a gun."

"Mama-*san*," Hara said, "I have an announcement: Emi-*chan* and I are getting married in June. We're going to Hawaii on our honeymoon."

"Congratulations," she said. "And just think, Hawaii! I can't believe you're going that far. It really is true what they say: this is an international age."

"Hey, Cho-*san*," Komaba said, "why aren't you going abroad?"

"Maybe I will," he said. "But someone has to look after your truck when you're gone."

"Do you want to buy it?"

"Perhaps. I'll have a look tomorrow."

"Oh, tomorrow," the old woman exclaimed. "You will want to go skiing. You'd better hurry off to bed."

We pulled our legs from the "fire" and padded off obediently to the next room, where seven bedrolls had been laid across the floor. A cookie was resting by each pillow, and a kerosene stove was aglow by the door.

The next morning when we arrived at the base of the mountain a little after ten o'clock, people were visible at every turn. According to a popular saying, every person in Japan is only one hour away from the sea and one hour away from the mountains. Nearly eighty percent of Japan is covered with mountains and virtually uninhabitable. Adding in the country's already limited land supply, the result is that over one hundred million Japanese live and work in an area smaller than the state of South Carolina. Not surprisingly, a Japanese ski slope on a Saturday morning in February is a product of this equation. Multicolored specks of pink and green seemed to appear out of nowhere and spiral down the mountain on top of one another until they gathered in a heap at the bottom like discarded confetti. Five separate chairlifts scooped skiers toward the summit, pausing at the lodge halfway to the top. To add excitement to this already frenetic scene, loudspeakers strapped to every other tree blared synthetic pop songs from the latest American teen heartthrob: "Electric Youth. Feel the power and the EH-NER-GEE."

At the top of the range our group divided. Cho, Hara, and Komaba took off down the most treacherous path, carrying a video camera and several blank tapes to record their exploits for posterity. The two women headed down a less difficult course, while Tomo and I — a beginner and a recovering cripple — chose the least rigorous route to the bottom. We agreed to meet at one o'clock for lunch.

Wearing borrowed pants that were too short and rented boots that were too tight, I soon realized that skiing down this mountain would be far more difficult than teaching English on crutches. Tomo and I managed to make it down the hill and up the lifts several times before Cho came skidding up to us at midslope and offered some much-needed instruction.

"How did you get to be so good at this?" I asked.

"I went to college," he said.

"To go skiing?"

"Not exclusively. We went hiking and bowling as well. Nowadays, my club even goes scuba diving in Okinawa."

"And you complain that you never get a vacation . . ."

"Now, I don't. I went skiing ten times a year in college, but I haven't been ten times since I graduated. I am supposed to have fourteen vacation days a year, but this is the first one I've taken. Being a teacher is like being a mother: we never have a day off."

The others had arrived for lunch before us and settled down in the cafeteria with matching bowls of beef curry over rice. I bought a large order and a small cup of tea, and took my place on the redwood bench.

"Excuse me," Emiko said to me as I started eating. "May I ask you a question?"

"Sure."

"It's about your country."

"Okay."

"My father owns a real estate company in Tokyo, and he has decided that now is a good time to invest in America. Last month he sent me and my younger sister to New York to examine some office buildings."

"Did you buy anything?"

"No. Well . . ." Her voice dipped down and lost its happy edge. "We looked at a lot of buildings, but the truth is, we had a lot of problems with our name."

"What name?"

"The name of our company."

"What is it?"

"Up River," she said. "Up River Real Estate Company."

"And how did you choose that name?"

"Because of Bridgestone Tire. Do you know this company? It was started by a man named Ishihashi. Ishihashi means 'stone bridge' in Japanese, so when he started his company, he named it Bridgestone."

The others nodded at the story.

"My father's name is Kawakami," she continued, "which means 'river up.' When he started an international company, he named it 'Up River.' "

"He sounds very smart," Chieko said.

"But every time we said our name, the Americans would laugh. Can you tell me why, Mr. Bruce?"

"I think so," I said. "The name of your company is very similar to the English expression 'up the river,' which means to be in prison. Some people may have thought that your name was a joke or that you had criminal connections."

"Oh no!" she shrieked. "I am so ashamed."

Like many others, the Kawakamis had decided that having international status begins with having an international-sounding name. English has such appeal as the lingua franca of modern life that the Japanese are labeling everything from toilet seats to hotel chains with exotic English terms. Supermarkets feature such goods as "Dish of Quickie" curry sauce, and restaurants welcome customers with slogans like "NICE EAT YOU?" In Sano I saw a pregnant woman with a T-shirt that said "WAY OUT, BUT CLASSIC"; and an advertisement in the train station advised new husbands to "carve a ham as if you were shaving the face of friend." All these foreign expressions, however, seemed less a sign of increased internationalization than an indication of how superficially most Japanese understand the West.

After lunch that day on the slopes, Cho and Komaba went off to the vending machines at the end of the room and returned with drinks for everyone.

"We bought a new type of beer," they announced. " 'The Winter's Tale.' "

"Oh, that sounds delicious," Chieko said. "I saw it advertised on TV."

"Why is it called 'The Winter's Tale'?" Emi asked.

"Because they make it only five months out of the year," Chieko said. "It's a *winter* beer."

"Maybe it's just colder," Emi suggested.

Hara suddenly interrupted. "Haven't you ever heard of Shakespeare?" he asked impatiently.

"Yes . . . ," Chieko said.

"Well, this beer is named after one of his plays. That's why there's a quote from him on the can."

"Oh," she said, "I didn't know."

"It doesn't really matter," Cho said. "The labels are only for looks, the beer is all the same." He opened up the can, took a sip, and splashed it around in his mouth. "It's just as I expected," he said. "The name may be from Shakespeare, but the beer is from Japan."

One of the complaints I heard most often in Japan was that young people, such as Cho and his friends, are becoming too Westernized. Unlike their parents, they have traveled abroad, honing their tastes in such exotic haunts as Surfer's Paradise in Australia and the Great Pyramids in Egypt. They are more accustomed to having money, and to spending it on conspicuous consumer goods like portable CD players and upscale, off-road pickup trucks. They are more apt to spurn formulaic marriages, and more inclined to take their unwed girlfriends for weekends of escape. Mr. C, in contrast, had never traveled abroad, had never been skiing, and rarely went anywhere with his wife. Yet if Cho and his friends are any indication, the reports of the death of Japanese culture are grossly exaggerated.

"Hey, little brother," Komaba called as soon as we slid our legs under the table for dinner, following our evening bath, "why don't you pour everyone some beer?"

"*Hai*," Tomo shouted, bounding to his feet and dashing off to the kitchen without comment.

"We have to train him to be obedient," Komaba explained. "He just joined the cheering club at his university in Osaka."

"I hear those clubs are very strict," Chieko said.

"It's true," Komaba said. "They meet every day and practice special mental training. They march around the campus screaming chants in the afternoon. Last year they even made a special trip to Tokyo to sing and pay their respects to the emperor."

"It sounds spooky," Emiko said. "It reminds me of the army."

"Maybe, but right now he is a freshman, so he is learning how to obey his elders. That's important for anybody to learn."

"Excuse me for being late," the boy barked as he came rushing back into the room. "I present you with beer." He made his way around the circle, pouring beer in each person's glass. When he finished, we raised our glasses and cheered: "*Kanpai.*"

"Hey, Tomo-*kun*," Emiko said as she sampled the sautéed scallops and pickled fish, "why did you join that club?"

"Oh, I don't know. I saw a similar group on television singing at a baseball game. I thought it looked interesting. They had a lot of group spirit."

"That's wonderful," she said. "So many students today don't think about the Japanese spirit. They just think about themselves."

"That's right," Chieko added. "We had a young teacher come to our school this year, and he didn't want to stay late or lock up the building, as new teachers are supposed to do. He only thought about what time he could leave."

"It's worse in Tokyo," Komaba said. "I had to interview people for the bank recently, and half of the students couldn't even speak proper Japanese. It was embarrassing."

We finished the appetizer, and the old woman set down a huge wooden boat of fresh sushi in the center of the table. "This is the life," Hara said. He plucked a sample from the tray, dragged it through a dish of soy sauce, and dropped it in one piece onto his tongue. "Most students don't understand this. All they eat is McDonald's hamburgers and Dunkin' Donuts."

"You people sound just like your parents," I said. "You even sound like *my* parents."

"But it's true," Cho said. "Japanese students just do what others around them are doing. If it's winter, they go skiing. If it's summer, they go diving."

"But you go skiing," I said, "and you'd go diving if you could."

"Yes," Emiko said, "but we are not *shinjinrui*." She spat out this last word as if the mere mention of it would poison her meal.

The others mumbled in assent. Komaba asked his brother to pour more beer.

"What's the difference between you and the 'new types'?" I asked.

"*Shinjinrui* are people who cannot think for themselves," Komaba said. "I went to buy a coat the other day in Tokyo, and all I could find were leather jackets. Two years ago it was football jackets from America; then it was navy pea coats. This year it's leather. The people who buy these coats are *shinjinrui*."

"The problem," Komaba said with a tone of authority, "is that with all the American movies and fashion clothes, students have forgotten what it means to be Japanese."

"Do you remember what Kana said the night of our *go-con*?" Hara asked.

"She said that Japan would soon rule the world," I said.

"I don't think it's true," he said, "but I think she made a good point. Our parents bowed down to Americans because they won the war. But since Americans became rich, they have forgotten how to work. I'm afraid that the same may be happening to Japan. The *shinjinrui* have lost the will to work. They think the world is all discos and dates."

"Hey, little brother," Komaba called, "why don't you show Bruce-*san* what you learned in your club? Show him what the Japanese spirit used to be like."

Tomo rose quickly to his feet and took his place in the center of the room. He stood still for a moment, his arms at his sides, his face stern with concentration. The girls laid down their chopsticks. The old woman peered out from behind the door.

Tomo bowed slowly to three sides. "*Hai, hai, hai,*" he grunted, drawing his hands into the air with razor-sharp precision and jabbing them into the room in rapid succession. "Now is the time," he cried, "push toward the line. *Banzai! Banzai! Banzai!*" He clapped his hands together three times, bent down, and with the sudden thrust of a catapult vaulted into the air, turned a somersault, and landed on his feet. The others gasped, then started to applaud. Tomo snapped to attention, bowed to the table, and returned to his seat.

"That's amazing," Emiko crooned. "He must have learned that in school. He sure is not a 'new type.'"

Laments about "new types" are quite old in Japan. In the 1860s, trendy young men were assailed for cutting off their topknots and eating Western beef. In the 1920s, young people (including Crown Prince Hirohito) were labeled *mobos* and *mogas* — "modern boys" and "modern girls" — for wearing bob cuts, carrying pocketbooks, and walking hand in hand with their lovers down the Ginza in Tokyo. In the 1950s, delinquent youngsters were scolded for listening to jazz and watching the ultimate token of Western lasciviousness, the kiss, depicted on the silver screen. Before the war, kissing had been banned in films and was edited out of imported offerings.

While each of these generations has been lambasted as heralding the end of Japanese culture, none in fact has brought its demise. Just as Japan's native Shinto creed made way for imported Buddhist temples and Christmas trees, so Japanese popular culture has made way for Swiss watches, American jazz, and now — gasp! — the French kiss. Even the idea of love, once considered a mortal threat to the Japanese family, has become as much a part of contemporary life as the arranged marriages of old. Each new generation of Japanese youth takes another step closer to the Western fold, but each one stops short of the line that would mean abandoning its identity.

After dinner we moved from our formal table back to the main lounge. Again we sprawled on the straw mats with our feet under the quilt. The Snow Country *mama* brought out cups of Earl Grey tea and plates of prepackaged cream cakes. Hara rose to replay the videotape he had filmed that afternoon on the slopes.

"Oh, you're soooo talented," the girls said as Hara came down the mountain.

"Oooh, *purofeshionalu*," they said when Cho appeared on the screen.

As I watched this group of young Japanese lolling around on tatami mats and watching a homemade videotape, I thought for a moment that they were close to achieving a balance between East and West. These people did not want to become Western, they just wanted to have a few more of the freedoms and trappings of Western lifestyle. Their faith in their country was profound, verging at times on arrogant, and they had no intention of forsaking the pledge they had made to Japan while they were still in school. This was, in the words of the T-shirt, the "way out, but classic" generation. They may hang out in big-city nightclubs and chase fairy tales of love, but they still feel most at home when their legs are under the *kotatsu* and their feet are over the fire.

14

BASEBALL, APPLE PIE, AND DRAGON MOTHERS: THE TEACHER IN JAPAN

To honor the teacher is a means of honoring the Way. Therefore, the teacher shall possess the justice that reigns between a ruler and his subject and the love that exists between a parent and his child.

— *Itō Jinsai, Japanese philosopher, 1666*

ONE MORNING IN EARLY FEBRUARY, I sat down with Denver to plan a game of English charades for his seventh-grade students. On a list of vocabulary words that students would be asked to act out in front of the class, I wrote the word "mother."

"What can the students do to act like a mother?" he asked.

"Oh, that's easy," I said, cradling my arms around a make-believe baby and pretending to croon a lullaby.

"But that won't work," he insisted. "That's not what Japanese students think about their mothers. Motherhood doesn't have the warm image in Japan that it has in America."

"Okay, what is the image of mothers?" I asked.

"*Kyōiku Mama*," he said. "The Education Mother."

The Japanese have borrowed baseball from the United States; they have lapped up apple pie as their own; yet they have stopped short of borrowing the American exaltation of motherhood. Mothers occupy a social position in Japan somewhat akin to that of teachers in the United States: they are essential for the welfare of the state, most people have fond memories of their own, but basically they are taken for granted, and certainly they are not lionized. The expression *Kyōiku Mama*, similar in tone to the term "stage mother," is used to describe women who pressure their children to study constantly in order to excel on standardized exams. Other nicknames for overbearing mothers include *Onibaba*, "Devil Woman," and *Mamagon*, "Dragon Mother."

After learning these terms from Denver and hearing them repeated all over school, I set out to learn why mothers earned such ignominy, and who took their place as the keepers of the flame. My first stop was the Cherry Blossoms' home.

About once a week while I was in Sano, I would visit Mr. C's home after school, spend the evening with his family, then drive with him the next morning to our office at the Board of Education. During these relaxed times, the evening pattern around the Cherry Blossoms' comfortable, two-story home was almost always the same: Mrs. C, a home economics teacher at a nearby junior high school, would prepare, serve, then clean up from dinner, and finally emerge from the kitchen around nine o'clock to straighten the house or do the laundry. Mr. C, who usually returned home after his wife, would rush through his meal, take a quick bath, and then do leftover office work on his personal computer in the den. Though they had been a "love match" in their youth, never once during the year did I see them make physical contact, share a flirtatious glance, or exchange more than passing words with each other.

"Would you like something to drink?" Mrs. C politely asked her husband after dinner on the night I came to talk about motherhood.

"After my bath," he said, standing up and unbuttoning his pants. "Then I'll have some sake."

"And Mr. Bruce?"

"The same," he answered. "And make it warm. The rice tonight was cold."

He dropped his pants and shirt on the floor and walked off toward the bath.

"My husband never waits for anything," Mrs. C said as soon as he had left. "He eats fast; he walks fast; he even speaks fast. Sometimes I think he knows only three words: *gohan, furo, futon*; food, bath, and bed." She picked up his clothes and laid them over a chair. "Do your mother and father talk to each other?" she asked.

"Yes," I said. "Especially before dinner. They call it 'cocktail hour.' "

"You mean they drink together, too? I can't believe it." She poured hot water from the teakettle into the sink and began to wash the dishes. "My father talked with my mother when I was young, so I thought this was typical. But my mother says no. She says my husband's character is normal for Japanese men. I wish I had married an American."

Japanese families, unlike those in the West, are not a haven for private love between individuals. Because most marriages in the past were formally arranged, the family has traditionally been seen as a functional social unit in which the husband earns the money and the wife tends the children. Japan has no ideal of Mom and Dad gathering the kids and the dog and heading out in the station wagon for a Sunday drive. Instead, the Japanese family is often lampooned as comprising an absent father, a nagging mother, and two children who go to school all day, attend cram classes all night, and rarely see their parents together.

Fathers have long been derided. According to popular lore, the four biggest fears of the Japanese are earthquakes, thunder, fire, and fathers. In the popular media, men are ridiculed as bumbling alcoholics who stumble in late from drinking parties every night and are wholly dependent on their wives when they are at home. "What is Boston Club Bourbon to you?" an announcer asks a typical Japanese man in one television commercial. "Boston Club is my wife, my son . . . my life," the man answers.

Women get even less respect. Despite the preeminence of

mothers within the walls of the family, where they control the money, the household, and the children, outside the home they are shunned, and often mocked by both their children and their husbands.

"All my wife thinks about is entrance exams," Mr. C said to me after his bath, as he poured me a cup of warm sake from a two-liter bottle with a rattlesnake coiled inside. "My older son, Yuji, must take exams next year for university, and Takuya for high school. All we talk about is these idiotic tests."

"But aren't they important?" I asked.

"Sure they're important. But children have other things to do. Sometimes I think that my wife knows only one word: *benkyō, benkyō, benkyō;* study, study, study." He offered me a plate of raw horse meat, which I politely declined. "It's very good," he said. "A delicacy. Anyway, what did your mother say to you?"

When I was a child, my mother nagged me with a different refrain, I told him. "Hobbies, hobbies, hobbies," she preached. "You must do more than homework; you must develop 'personal interests' as well."

Mr. C slammed his cup on the table, clapped his hands twice, and bowed his head as he did at the Shinto shrine. "You don't know how lucky you are," he said. "Japanese mothers never say that to their children. All they push is homework. Learning is not important, only studying."

Several days later I broached this subject with a group of young male teachers at a Denny's family restaurant. It was late on Friday night, and we had stopped off after an informal drinking party for a late-night bowl of *rāmen* noodles.

"I am really worried about my students," said Machida-*sensei*, the stylish math teacher who had objected when the principal told him to wear a tie to class. "They seem to have no character. When they go home in the afternoon they should be doing warm things like reading books or playing sports. Instead, they are always studying. They get this from their mothers."

"Last night a mother called me at eleven P.M. to talk about next week's exams," complained Hongo-*sensei*, a physical education instructor. "I couldn't believe it — eleven o'clock at night! I have no privacy anymore. I never get any sleep. For the students it must be even worse. They have to live with these people."

"Is she a *Kyōiku Mama*?" I asked.

As soon as he heard this term, Hongo-*sensei* jumped to his feet, wrapped a napkin around his head, and pretended to draw a sword from his waist. "*Kyōiku Mamas* beware!" he sneered. "We know where you live." He thrust his imaginary saber into the air with a snarl. Several couples in the restaurant glanced over at our table, and Hongo-*sensei* returned to his seat with one final riposte.

"I think the problem is that parents don't like to teach their own children," Machida-*sensei* said when the saber-rattling was done. "Even if it's using chopsticks or getting dressed, they expect the school to do everything. Several mothers even called the principal over the New Year's holiday and complained that we were not giving students enough to do."

"You've got to be kidding," I said.

"It's true," Hongo-*sensei* added. "The principal called me on the telephone and told me to start basketball practice a week before classes began. That's crazy."

"Would this happen in America?" Denver asked. "How much time do teachers spend in school?"

"Not as much as you do," I said. "American teachers seldom come to school on weekends, and almost never during vacation. Of course they have a lot of work to do, and often take papers home, but most would never put up with this."

"Do they have to make home visits?" Hongo-*sensei* asked. "What's that?"

"We have to go to every one of our students' homes at least once a term," he said. "Next week I have to visit forty-five houses in three nights and write a report on each one. The report has to say how much time students study, how much television they watch, and what their rooms look like. The principal makes us do it. He thinks we are the education police."

"We don't do that in America," I said. "Parents sometimes come to the school, but teachers rarely visit homes."

"You see what I mean," Machida-*sensei* said. "We have to do too much. We hardly have time to teach."

The teacher in Japan has long been accorded a special, almost sacred status. In a country that views schools as secular cathedrals, teachers have become lay priests. The word *sensei*, though commonly translated as "teacher," in truth has no equivalent in English. The two Chinese characters that make up the word literally mean "one who was born before." The essential ingredient for a *sensei* is the wisdom he or she has gained through experience, not through reading books. In Japan, the wise one learns through time. Even today, the use of the word *sensei* as an honorary appendage to names is not limited to schoolteachers alone. Any valued adviser or mentor can earn the respect inherent in the word *sensei*.

In premodern Japan, schools were built around the personality of an esteemed instructor. The Tokugawa shoguns who ruled the country between 1603 and 1868 established schools in many regions to train bureaucrats to run the state. In these Confucian schools the master-disciple relationship was central: instructors led through the example of their own character and conduct. "The teacher," proclaimed one of the most widely used textbooks at the time, "is like the sun and the moon." In many cases the students moved to these schools and ate, slept, and even bathed with their *sensei*.

In the years leading up to Japan's war effort in the 1930s and 1940s, military discipline became even more central to schools, and teachers were required to undergo martial training themselves. Eventually military officers were assigned to schools to work alongside teachers. The old master-disciple relationship was not abandoned but rather was co-opted by the state. Teachers were still expected to lead by example, which in the new nationalistic context meant showing the utmost respect for the emperor. A "Memorandum for Elementary School Teachers" of the era advised: "Loyalty to the Imperial House, love of country, filial piety toward parents, respect for superiors, and charity toward inferiors constitute the Great Part of human morality. The teacher must himself be a model of these virtues in his daily life, and must endeavor to stimulate his pupils along the path of virtue."

When the American education authorities examined Japanese schools after the war, they vowed to eliminate these "authoritarian ideas" and replace them with "democratic values." One of their initiatives was to encourage teachers to form unions. Within several years the newly formed Japan Teachers Union (JTU) boasted membership of almost eighty-five percent of all teachers in the country.

Much of postwar education history in Japan has been dominated by a struggle between the conservative national government, led by the ruling Liberal Democratic Party, and the liberal teachers' union, controlled by the Socialist and Communist parties. While the JTU has been a vocal proponent of reduced federal control in education (as the Americans had hoped) and has gained some limited victories, it has been unable to withstand the monolithic pressure of Mombushō, Japan's Ministry of Education. From JTU's peak in the late 1940s, both its power and its membership have declined precipitously in recent years. In Sano and southeast Tochigi, the JTU had no presence at all, and the teachers I worked with were affiliated with another, less vocal union.

Teachers often complain that they have little freedom over what material to teach. With curricula written and approved in Tokyo, all classes follow predetermined schedules which ensure that all students in Japan study the same material at roughly the same time. Just as Napoleon censored school textbooks to stress the state over the individual, so the Japanese government strictly controls what information arrives on students' desks. This oversight is so extensive that in the case of junior high school English, the Ministry of Education publishes a list of 350 English words — from "a" to "young" — that all students are required to know before graduation. Yet despite the solid state control of classroom content, teachers still feel responsible for the lives of their students. In the week leading up to our Friday night conversation, Machida-*sensei* had to be called away from class twice to retrieve a student who had returned home during the day, and Denver had to cancel dinner plans with me because of a special meeting with the principal to handle an incident in which a student had been caught drinking at home by a neighbor.

Compared with the United States — and most European countries as well — Japan has an essentially homogeneous culture, with a common moral and religious heritage. Parents are more willing to give schools the authority to teach their children the common "Japanese" values of hard work, self-sacrifice, and national pride. Teachers, the ones who assume this burden, are thus given responsibilities that stretch far beyond their classroom door. As Machida-*sensei* said after he retrieved his student from playing hooky, "If I don't get him now, who will? If I don't help him today, who can?" This type of teacher, one who takes responsibility for the personal development of his students, who not only teaches science by day but coaches tennis in the afternoon and makes house calls at night, is lauded in Japan as a *Nekketsu Sensei*, roughly translated as a "Hot-Blooded Teacher." In a straight popularity vote, the *Nekketsu Sensei* would outpoll the *Kyōiku Mama* by a margin of ten to one.

On the Monday morning after the gathering at Denny's I went to see Mrs. Negishi, who in addition to being the teacher of forty-five ninth-grade students was the mother of two preschool boys. During a break between classes she leaned on her desk and told me the story of why she became a teacher.

"When I was a junior high school student outside Tokyo, a soldier from America came to my school. The soldier was tall, with bright red hair and a shiny blue uniform. When he came to our class, he spoke too quickly for me to comprehend. I wanted to speak to that man, but I was afraid I could not catch what he was saying. I was very shy.

"After class, I spied the soldier walking out the back gate. I ran over to him, panting and out of breath, and uttered only one word, 'Where?' At first he looked at me for a moment, then he pointed out the gate and said some words that were too fast for me to understand. But that didn't matter at the time. I was so happy that this foreigner — this big, important man in a uniform — could understand me, that right there, standing in the middle of the school yard, I began to cry."

She smiled, fighting back tears again, and ran her hand across a photograph of her homeroom class that she kept on her desk.

"I want my students to have the feeling that someone important understands them. It doesn't matter what language, as long as they know that someone cares — not about tests, or grades, or colleges, but about them. In my class, I try to do that."

The third-period bell sounded in the middle of our conversation. Students and teachers flooded into the office. Mrs. Negishi motioned for a tall boy standing at the door to join our conversation.

"This is Sugiyama-*kun*," she said. "He is one of my best students, and next year he is going to attend the most prestigious high school in Tochigi." The boy blushed. "He has been having some trouble with English, so last week I went to his house to help him prepare for the test. I am sure that he will do well."

I wished the boy good luck.

"Thank you very much," he said. "I'll do my best." He bowed and scampered out the door.

The profound attachment between teachers and students is the main reason why teaching remains a popular profession in Japan. Especially in small communities like Sano, teachers have genuine stature in the community. But sadly, intangibles like respect from parents and love for children have become the last job benefits to attract young people into education. As in the West, more and more people in recent years have been turning to the more lucrative and "exciting" careers of international business and finance. Although Japanese teachers earn high marks for their community service, they also must work painfully long hours, teach in overcrowded classrooms, and earn low wages. The starting salary for a university graduate like Denver with a comprehensive teaching certificate and three years of training in the system was roughly fifteen thousand dollars a year, before national and prefectural taxes and a monthly deduction for school lunch. A teacher like Mrs. Negishi, with over fifteen years' experience, earned less than twice that amount. All this is true in a country where the cost of living is significantly higher than it is in the United States.

As much as she loved her job, Mrs. Negishi regretted that it left her little time for her husband and her children. She was a doting mother and often showed me photographs she had taken of her two boys as well as pictures they had drawn. In a show of professional prudence, however, she kept these pictures hidden in her desk.

As we walked to the fourth-period class after our talk about teaching, Mrs. Negishi seemed unusually distracted.

"Is something wrong?" I asked.

"Oh, nothing," she said.

"How are your boys?" I asked, suspecting a problem.

"One of them has a cold today, so he is staying at home. I couldn't be there with him, so I had to ask my mother to come and stay for the day."

"I'm sorry," I said. "I hope he gets well soon."

"This morning, as I was leaving for school, my son said to me, 'Mommy, which do you love more, your students or me?' " She paused before entering the classroom and stared down the empty hall. "Of course I love my children," she said, "but I spend so much time at school."

In the seventh-grade game of English charades which had launched my search, a young girl drew the word "mother." She walked to the center of the circle, raised her hands to her head like a pair of horns, and wagged her finger in front of her face in a nagging, menacing way. In no time the students had guessed the word.

"When mothers get mad," the girl explained, "they become like the devil, and all children know that the devil has horns."

After several more rounds of charades, a boy drew the word "teacher." He marched to the blackboard and began writing furiously in the air. Without stopping for an answer, he sat on the floor and started scrawling on an imaginary pad. Finally, he pretended to take a mop and run it across the floor. Writing on the board, scribbling at a desk, cleaning the classroom floor — again the students had no difficulty guessing the word.

After class I asked these seventh graders who they thought did the most to prepare them for everyday life. The results were overwhelming: one student said his father, six said their mother, and the rest of the class — thirty-five students — chose their teacher.

While mothers remain at home, pushing their children to study hard for exams, teachers take over at school, mothering their students to work with others and develop strong moral values. This arrangement breeds tension between parents and teachers, who often have different goals. A well-known expression in Japan warns, "Any nail that sticks up must be hammered down." This means that any student who shows exceptional ability must be muted to fit in with the group. In the classroom, students are taught not to flaunt their talents, "but," Denver explained, "mothers want their children to succeed, to earn merit — to be protruding nails." The *Kyōiku Mama* is born of this system.

To be sure, Japanese students still love their mothers. "Just remember," Denver said, "mothers are still mothers. Japanese boys may complain a lot, but they always go crying on mama's shoulder." The difference between Japanese and Americans, he said, is that Japanese children don't think of their mothers as the apple of their eye. To make his point, Denver told me a story about an American football game he had seen on television. One of the players had worn a headband that said "HI, MOM."

"I love my mother, too," he explained, "but I could never wear *that*. I would be too shy."

Although Japanese sumo wrestlers are not likely to shout "Hi, *sensei*!" into waiting television cameras anytime soon, perhaps that gesture would come close to capturing what students feel toward their teachers. "While my chess-loving father failed even to entertain me," Natsume Soseki wrote in his famous novel *Kokoro*, "*Sensei* gave me far greater intellectual and spiritual satisfaction as a companion . . . Indeed, it would not have seemed to me then an exaggeration to say that *sensei's* strength had entered my body, and that his very life was flowing through my veins. And when I discovered that such were my true feelings toward these two men, I was shocked. For was I not of my father's flesh?"

Parents may provide the flesh and blood, but teachers provide the powerful example of their own commitment to serving the state. In our discussion after class, Denver put it best when he told me that the prevailing icons in Japan are not baseball, apple pie, and motherhood but *yakyu, miso shiro,* and *Nekketsu Sensei* — baseball, miso soup, and the Hot-Blooded, Warm-Hearted Teacher.

15

LEARNING TO CRAM: THE JUKU GENERATION

I memorize French declensions until midnight.
I wonder, what's the name of that insect?
Shouldn't I be able to memorize more easily?
Half of my life — blank spaces in this homework.
— *Anzai Hitoshi, "Homework," 1960*

"ARE YOU READY?" *Hiroyuki Arakawa called out in halting English. An actress straightened her skirt on the fringe of the makeshift stage. A few stragglers whispering in the corner turned their eyes toward the center of the room.*

"Take one," he cried, drawing his arms together with a crisp snap. "Begin!"

Kumiko Yamaguchi walked slowly across the bare floor in oversize slippers and stopped just shy of a man seated alone on a bench. The Wednesday afternoon sun cast its last shadow across his face.

"Excuse me," she said. "What time is it now, please?"

"It's six o'clock," the man answered.

"Thank you very much." She turned quietly and retraced her steps.

"Cut," Hiroyuki cried. "Let's try that again . . ."

Hiroyuki Arakawa has never directed a film. In fact, he had seen only one movie in his life before he walked onto this set. Hiroyuki was nine years old and a fourth-grade student at Sano Elementary School. Although he would not begin formally studying English until seventh grade, Hiroyuki learned this director's shtick and several other routines at a private English academy he attended twice a week after his regular school day was over. For one hour every Monday and Wednesday afternoon, Hiroyuki joined his eleven-year-old sister and three neighbors to practice the fundamentals of Living English at an after-hours cram school, or *juku*.

While others might be playing baseball or watching TV, Hiroyuki learned some basic English questions ("What day is it today, please?"), some simple responses ("Today is Wednesday"), and the numbers from one to one hundred. Hiroyuki is part of a new breed of Japanese youth who spend their childhoods in preparation for the double jeopardy of high school and university entrance exams. He is a member of the *juku* generation.

Hiroyuki's teacher was a thirty-two-year-old former sushi chef named Nobu Ishikawa, who was a high school acquaintance of Cho's. An unusually tall and lanky man with large hands, a pointed nose, and round, expressive eyes, Ishikawa learned English through years of reading grammar manuals and collecting conversational guidebooks. His private English *juku*, which he nicknamed SPEL, for Sano Preparatory English Lesson, was his sole source of income. His classes ranged from "Getting to Know English" sessions for elementary school students to rigorous exam-cramming courses for anxious high school seniors preparing for the February exams. The proliferation of this type of school represents one of Japan's fastest-growing industries in recent years — the business of English.

"To master English is very important, especially for Japanese," Ishikawa said between classes in his stark two-room schoolhouse above an abandoned garage. "Japanese students have entrance exams; Japanese companies do business abroad. I even have some students who are over sixty years old and are studying English because they can't read the labels in a 7-Eleven. It is not necessary for you to learn Japanese, but it is necessary for us to learn English."

Compared with the English spoken by most junior high school English teachers, Ishikawa's speech flowed smoothly and without hesitation. His pronunciation was clear, and his style relaxed and colloquial.

"At my school we have a philosophy," he said, using one of the many English expressions he liked to sprinkle in his speech: "Slow and steady wins the race."

The modern *juku* arose from a century-old tradition of preparing students for higher education. In feudal Japan, a person's rank, family, or class determined the path that he or she would follow through life. The sons of samurai had the opportunity to land prestigious jobs with the state; the sons of peasants did not. With the elimination of the caste system by the Meiji emperor in the late nineteenth century, education replaced class as the new stairway to success. In 1877 the government established the Imperial University of Tokyo, Todai for short, which quickly became the ultimate goal for students who wished to join the civil service or become business executives. Even after the government set up five more universities around the turn of the century, competition for slots was fierce. In response, government officials were forced to devise a demanding selection process. Following the Confucian tradition, they chose exams as the best measure of success.

Since its inception, the university selection process has been pivotal because of the importance placed on graduating from a select institution. In 1937, for example, seventy-five percent of the people accepted for upper-level civil servant jobs and almost half of company presidents came from Todai. In 1987, the figures were almost identical. This connection between jobs and schools has not been lost on ambitious mothers, who explain to their children that in order to achieve a prestigious career, they must attend an elite university; in order to attend an elite university, they must pass that university's entrance exams; and in order to pass the exams, they must begin preparation at an early age. Thus, even as the number of publicly funded schools increased in the twentieth century, so did the number of *private* academies, which — for a fee — promised to prepare students for the exams.

If anything, the plight of students has gotten worse in the wake of the postwar educational "reform." One of the principal tenets that the American authorities introduced into the Japanese schools during the Occupation was the idea of "total equality" based on merit, which they viewed as a means of guaranteeing democracy and avoiding the elitist domination of prewar Japan. The Allies abolished the imperial university system and opened up higher education to a greater number of people. The number of universities in Japan surged from 48 in 1945 to 201 in 1950, and continued growing to reach 500 by 1980. The number of students in these universities increased accordingly, from three percent of high school graduates after the war to thirty-seven percent today. To make this system more meritocratic, the Americans insisted that candidates for admission to universities earn no special points for a stunning letter of recommendation, a powerful backhand, or a talent for the oboe. Only test scores would count.

Unwittingly, this plan pushed competition even lower down the age scale. Under the new system, access to competitive high schools would have to be determined by additional entrance exams at the end of junior high school. These days, any student who realistically hopes to attend a first-rate university must not only pass that school's exam at age eighteen but also pass an exam to enter an academic high school at age fifteen. As a result, seventy-five percent of all students attend some kind of cram school.

The entrance exams in Japan are unlike any tests in the West. First, they are not required for graduation from a particular school — like the "O level" in England or the *baccalauréat* in France — but are used solely for admission to a higher-level school. In addition, unlike the SAT in the United States, one test does not work for all institutions: different schools use different exams. Finally, all universities hold their exams at roughly the same time, so that most students can take only one test at a time. If an applicant fails, he or she must wait an entire year for another chance. About fifty percent of students do fail the exams for the college of their choice and take an additional year to study *exclusively* for these tests. Such students are called *rōnin*, a name once given to samurai who were cut off from their masters and forced to roam the countryside. An average of three quarters of all students admitted to the University of Tokyo have been *rōnin* for at least one year, and some have done nothing but study for the Todai exams for three or four years.

The reason for this high rate of failure is the nature of the entrance exams. At the university level, the tests are complex trials that require multiple sittings on successive days and cover a wide range of subjects from the theories of ancient Greek philosophers to the lengths of major rivers in Africa. At the high school level, the entrance exams take one day and contain sections on most subjects taught in junior high school, from Japanese to mathematics, history to science. Although the material changes from year to year, one part continues to give students the most difficulty. Every exam, for both high school and university, contains a section on English.

"Good evening," Ishikawa said to the five high school boys trudging into the SPEL classroom at seven-thirty that night. "How are you today?"

"Tired," one boy grunted in Japanese as he slumped into a seat in the first row. "I can't believe how much work I have to do. I have a test tomorrow and a basketball game on Friday night."

"Let's begin our English class for today."

There are basically two types of cram schools: huge chain stores that pack students by the dozens or even hundreds onto long benches where they watch video lectures about English grammar and sentence structure, and smaller, more intimate academies where one teacher works closely with a limited number of students. The larger houses often have computers to evaluate students' test-taking skills, while the smaller ones offer individual attention by a teacher who can spend more time with each pupil. Their advantages are different, but their objectives are the same: to help students through the teenage rite of passage known as *juken jigoku*, "examination hell."

"Tonight," Ishikawa continued, "I have prepared a short vocabulary test for you to take. This is material you should have studied for today. You have fifteen minutes. Please begin now."

The test consisted of twenty-one English words; students were asked to supply the officially sanctioned antonym for each one. *Gallant. Righteous. Sagacious. Frugal. Friendly. Figurative. Fantastic* . . . The items came directly from the approved list of five thousand words that all graduating high school students are required to know in order to pass the English portion of most university entrance exams. Silence descended as the severity of the assignment sunk in. Ishikawa graded some papers at his desk; the students squirmed in their seats. *Painful. Destruction. Consumption. Pessimist. Utmost. Treachery* . . . At the quarter-hour mark the test ended. A quick survey of the room revealed that none of the seven students had successfully completed more than five pairs. The teacher was concerned.

"I see you have not studied hard this week," Ishikawa said, glaring down at the boys. "We have a lot of work to do. An ounce of prevention is worth a pound of cure."

As the students began their review, I soon realized that none of these boys would be angling for a slot at the University of Tokyo. Indeed, none of them would be heading outside Tochigi after high school. Yet each one still willingly paid more than $240 a month to study English twice a week. At the end of the hour, I got a chance to ask them why.

Mitsutoshi Iwasaki, the son of a taxi driver, explained. "Everyone else was going to *juku*," he said, "so I thought I should go, too. It's a little expensive, but my father said he would pay. I want to study computers in university. I don't want to drive a taxi."

Yoshiro Kobayashi's father harvested bamboo for a living. "I helped him out a few times," said Yoshiro, a frail boy with glasses, "but I am not strong enough to do that kind of work. Plus I don't like it all that much. I want to become a train engineer. My mother suggested that I come here, and I agreed."

While these students had no hope of entering a high government agency or major international bank, they did hope to improve their future with white-collar jobs that paid well. For them, as for the brightest students, the key to these dreams lay in the *juku* system.

"If you work hard," Ishikawa reminded his pupils at the end of class, after they had reviewed their vocabulary and interrogative pronouns, "you can make your dreams come true. I want you to spend more time with your exam books this week. Next Monday we will take this test again. Remember, where there's a will there's a way."

At nine P.M. Ishikawa-*sensei* welcomed seven ninth-grade girls and boys into his three-table classroom. Most of the students were still dressed in their formal black uniforms. They carried their school books in their government-issue bags. In an effort to coordinate his lesson with the school's, Ishikawa-*sensei* taught from the same textbook that the students used in English class. Like many old-fashioned teachers, he began with recitation practice. Standing behind his desk, he read through Lesson Ten, "Two Visitors," and recorded his time on the blackboard: three minutes and forty-seven seconds. The students then took turns reading the chapter out loud, and anyone who completed the four pages within the target time period was given a lemon drop.

Pavlov would have been proud. The rest of the hour was spent reviewing, question by question, the midterm test administered that day at Sano Junior High.

"How did you find the test?" Ishikawa asked as one of the students handed him a mimeographed sheet.

"Too difficult," the students mumbled.

"How about the vocabulary?"

"Too many words."

"And the written part?"

"We couldn't understand."

"Let's have a look." He turned to the last page of the test and began reading a sample question:

> Do you know the original meaning of "rival"? It was a man living by a river or a man using the same river with another. Rivers were very important, and people living by a river came to compete with each other about rivers — when they caught fish and when they used the water of a river. In this way, the meaning of "rival" became a man who competes with another.

The students groaned as he made his way through a set of questions designed to test their comprehension of the passage. What is the original meaning of the word "rival"? How did the current meaning of "rival" develop from the original meaning? When do we use "rival" today? As he read, the students grew more agitated.

"Who made this silly test?" one of the girls blared out.

"Fuji-*sensei* made it," said a boy with half of his shirt buttons undone. "I hate him. I bet I scored only fifty percent."

"*Sensei*, why do teachers make such hard tests?" the first girl asked.

"Maybe because they want you to work harder," Ishikawa suggested.

"But I never get good marks in school," she continued. "I can't understand what my teacher says. He always wants us to memorize the words and repeat after him, but I can't understand what he says." Her voice trailed off to a whine. "I hate school. I don't learn anything in English class. That's why I have to come here."

One of the greatest ironies of Japanese education is that because schools devote so much time and energy to teaching "warm" things like moral education, national pride, and group cooperation, teachers often have little time to give proper instruction in their own subjects. And since the *sensei* are busy mothering their students, the *juku* master must teach them the basic material for the tests. Part of the problem facing classroom teachers is the vast range in students' abilities. Since schools do not group students by aptitude, teachers have difficulty adjusting the level of their lectures. According to one popular quip, the Japanese schools are like *Shichi-Go-San* — a holiday when children aged seven, five, and three are blessed at Shinto shrines — because the number of students who understand their teachers is seventy percent in elementary school, fifty percent in junior high school, and thirty percent in high school. The Japanese education system, for all its achievements, regularly fails to serve the educational needs and bring out the potential of a significant portion of its student population. One of the greatest strengths of the system — its ability to foster a communal environment among a wide cross section of students — breeds one of its biggest drawbacks — the failure of schools to prepare students for future examinations.

This impasse raises some troubling questions. Why are schools not adequately training students for the exams that the government has decided should measure students' abilities and determine their futures? Does the problem lie with the classroom curriculum, or with the tests? Should the system be changed?

Efforts to find answers reveal one of the founding tenets of modern Japan: schools are designed primarily to serve the needs of the state and the companies on which the nation relies. Just as the government tightened school dress codes in the 1970s because universities and companies complained of a loss of discipline, so it continues to put children through "examination hell" because the same powerful institutions claim that the tests are effective screening devices. Because most students "play" when they are in college, corporations cannot rely on college transcripts as a gauge of ability. Therefore, they choose employees based almost exclusively on the universities they attended, thereby increasing the importance of the exams that the students had to pass in order to be admitted.

The intimate relationship between big business and education came sharply into focus during my tenure as a teacher when a major political scandal erupted in Tokyo. The scandal, which resulted in the resignation of one prime minister and the arrest of several corporate executives, revolved around the question of who would be placed on select government committees that set guidelines for entrance exams and corporate recruiting. The main culprit in the affair, the Recruit Company, attempted to bribe politicians with shares of unlisted stock to ensure that its president had a place at the table where the rules were written. It is telling that Japan's biggest scandal of the last twenty years revolved not around violence, drugs, or even sex, but around education — specifically, the exact date on which university seniors would be allowed to have their first interviews with prospective employers.

While business leaders endorse the exams, parents and children alike despair at the amount of time required to prepare for them. The consensus is overwhelming: eighty percent of parents say the trend in *juku* attendance has gone too far; ninety percent of high school students planning to attend university say that they are tired of studying for exams; and in one poll, half of the high school graduates who were studying exclusively for entrance exams said that what they wanted most was "to go to a faraway place and drop out of sight from others."

Teachers also feel that *juku* are unnecessary. "*Juku* is a waste of time and money," Mrs. Negishi told me. "So many students study the book in *juku* that when we reach that chapter in class, they get bored and fall asleep."

Denver, who like many younger teachers attended *juku* himself, pointed out that some students enjoyed going to these classes with their friends, but that they probably were not necessary. "Students don't need *juku*," he said, "their parents do."

Still, this machine continues to run on the faith that the system, for all its flaws, offers the most tested route to success. If education has done anything for Japan in the last half century, it has opened the doors of social mobility. The key to the Japanese postwar economic "miracle" has been the millions of students who walked through the painful corridors of "examination hell" hoping for a chance to pass through the open door at the other end onto a successful career path.

"The *juku* is the key to college," Ishikawa said to me after the last students had left SPEL for the night just after ten-fifteen, "and college is the key to success. In Japan, if we graduate from a university we are treated with respect. We can get a good job, lots of yen, and enjoy a stable lifestyle." He picked up the junior high school tests strewn across his desk. "But studying in public schools is not enough."

"Why not?"

"Because the schools pretend that all students are created equal. This may be good for their hearts, but it is bad for their minds. Some students need special attention. How do you say it? Tender Loving Care."

"But there's one thing I don't understand," I said. "At school, the teachers all tell me they know which high schools the students will attend even before they take the tests. Why can't they eliminate at least the *high school* tests altogether?"

"They probably could," he admitted, "but *juku* would still not go away. Think of the students who come here every night. They know that the only way to get ahead in life is to work hard toward a specific goal. I like to tell my students what happened to me. I was a *sashimi bōchō*; I could use a knife to prepare fish in the traditional way. Not many modern Japanese people can cut fish properly with a knife. But in the company where I was working, I met the 'big wall.' I knew I couldn't move any higher. So I left, and started my own school. I had always wanted to be a teacher, and this way I can earn money, too."

He glanced around the empty room. Stacks of practice exams covered the shelves. Lemon-drop wrappers littered the floor. Discarded slippers lay huddled by the door.

"I want to help my students make their dreams come true," he said. "Little nine-year-old Hiroyuki told me the other day that he wants to visit foreign countries when he grows up. Yoshiro wants to drive trains. They all have dreams, and I can help them. But I can't take the tests for them, so I teach them what I have learned."

He turned in his chair and pointed to a sign taped above his desk. In simple, handwritten English letters, Ishikawa-*sensei* had summarized his formula for surviving "examination hell" and his advice to the dreamers of the *juku* generation: "NO PAIN, NO GAIN."

16

CLIMBING THE LADDER: DRINKING ALONE IN RURAL JAPAN

He had a dream, and behold a ladder was set upon the earth, and the top of it reached to the sky; and behold the angels of the Lord were ascending and descending on it.

— Genesis 28:12

THE CHESTNUT TREES were bare against the granite sky and the puddles were frozen like mirrors on the ground when I left my home on a Saturday night in late February and ventured into town. I met my American friend and fellow schoolteacher Jane at the Sano train station at seven o'clock for what we had agreed would be a study session on the Japanese male pastime of *nanpa*, picking up girls. In this modern dating ritual, Japanese boys cruise the bars around town, proposition girls with discreet queries (for example, "Would you like to drink some tea?"), and then spirit them into the night. After "drinking tea" and "eating rice cakes" the boy drives the girl home and bids her good night, and everyone lives happily ever after. Since my fairy tale *go-con* had ended unhappily, I thought I might fare better in the woods of *nanpa*.

In the United States, the preferred method of moving from one drinking establishment to the next is hopping. In Great Britain, one crawls. In Japan, perhaps as a result of its hierarchical nature, this activity is achieved by climbing. The nocturnal sport that Americans call bar-hopping and Britons call pub-crawling, the Japanese refer to as *hashigozake o yaru*, climbing the liquor ladder.

We began our quest in a traditional Japanese-style bar and grill called The Brothers, where customers gather alongside the counter before a large glass case of skewered chicken parts and vegetables, to drink, chat with the master, and eat *yakitori*, Japanese shish kebab. After perching ourselves at the bar, we ordered two Asahi Super Dry beers, several sticks of the house specialty — grilled chicken chunks and spring onion slices dunked in a tangy dressing of soy sauce and sugar — plus "Italian" tofu steaks with stewed tomatoes and bell peppers.

The room whirred with the sounds of an early evening crowd. Several groups of businessmen were crooning in the corners, and a small group of women sat quietly at an upright table in the center sipping Irish coffee. A middle-aged man seated alone next to me solemnly cradled two beer mugs and discussed the wonders of sumo wrestling with the master behind the bar, who was so busy basting chicken that he had little time to listen. After several minutes a much younger man came skipping over to the bar from the other side of the room, slapped the old man on the shoulders, and said, "Hey, buddy, what are you doing here? Are you doing *nanpa* again?"

The question startled us from our tofu steaks. Had we hit the jackpot so soon? I turned toward the two men. "Excuse me," I asked in the reserved tone of Japanese I saved for my most abrupt questions, "are you really doing *nanpa*?"

"*Nanpa*?" the man rejoined. "Do you know *nanpa*?"

"No," I said, "but I'm trying to learn. Will you teach me?"

The younger man, somewhat embarrassed, assured me that he didn't do *nanpa*. Only young boys partook in such folly, he said. Then, before even telling us his name, he returned my blunt question with one of his own: Would I teach his four-year-old son to speak English?

"I'll do anything for you," he promised. "I'll invite you to my house. I'll take you to Tokyo. I'll even give you a portrait of your dog."

"A portrait of my dog?" I repeated.

"Or your cat."

Hisashi, a twenty-eight-year-old man with thinning hair, droopy eyelids, and puffy, Charlie Brown cheeks, introduced himself as the leading seller of professional dog portraits in all of Japan. He was, it turned out, the only seller of professional dog portraits in all of Japan. A thriving art magnate of sorts, he lived with his wife and two sons in Sano, where he was slowly amassing a fortune and cornering the market in pet portraiture.

"I want to be very rich," he said with conviction. "I want to own a lot of land, a big house, and a Benz."

His business worked very simply, he explained. Every month he placed an advertisement in the leading — and only — dog owners' magazine in Japan. Zealous canine owners across the country sent him photographs of their pets, which he forwarded to an artist in Hong Kong. A month later he received a beautiful likeness of the pup, which he then sold to the customer at a five hundred percent markup. He worked by himself. He had never met the artist; he had never met a single customer. He didn't even own a dog.

"Won't the market eventually run dry?" I asked.

"No problem," he answered with all the aplomb of a seasoned entrepreneur. "Dogs die every day."

The doorman at the Magic Fish welcomed Jane, Hisashi, whom we had invited along, and me with a bow and led our party to the second-floor dining room, which had tender tatami floors underfoot and exposed cedar beams overhead. A rope hung from the ceiling and supported a two-foot wooden carp with a caldron dangling from its mouth over an open fire. The room glowed in the warm flush of green and white paper lanterns. We folded our legs under a table and ordered a round of Lemon Highs — a potent mixture of rice wine and lemonade served in a beer mug over ice — and several snacks, including boiled squid, raw shrimp, and salted lima beans. The conversation took another unexpected turn as Hisashi mentioned in passing that he had lived in France for four years.

"Why did you go to France?" Jane asked.

"Because my parents have money and they said, 'Go to France.' But I hated every day. I was lonely and sad and didn't speak any French."

While forlorn in Paris, Hisashi learned a lot about Japan. He discovered that Japanese goods, like cars, radios, and VCRs, were sold all over Europe. "I didn't think the Japanese were rich like this," he said. "I thought we were very poor because we couldn't speak English or use a knife and fork. I thought we were like potatoes." In Japanese slang an *imo*, or potato, is a country bumpkin. "But in France they thought we were rich."

After moving back to Japan and taking a job with a large construction company in Tokyo, Hisashi found himself with a small apartment, a girlfriend, and a bleak future, so he opted for life as a potato. He quit his job, moved back to the country, borrowed money from his parents, and began his art business. "To be happy," he declared, "you need money."

His quest for money and his wealthy roots propelled Hisashi toward his nontraditional career. "But," he insisted, "there are many people like me. I have a wife, a child, a small house, and a car. In a sense, every Japanese is like me . . ."

He took a sip of his drink and plopped some beans into his mouth.

"Originally," he admitted, "it was difficult. My friends said, 'Why don't you just join a nice company?' My mother said, 'You should do what other people do.' But I wanted to be on my own."

Although he enjoyed the rewards of working at home, close to his family, Hisashi still missed the camaraderie of a company. He had no drinking parties with colleagues after work, went on no company excursions, and attended no "forget the year" parties at New Year's time. His only friends, he said, were his junior high school classmates.

As he talked Hisashi grew gradually more disconsolate. Then suddenly he glanced at his watch and mumbled something about a party of old friends. He apologized, threw down some money on the table, and said he was late for a game of mahjong. "If I don't go now, I never can," he said as he left. "If I don't conform some, they'll think I'm strange."

We left the restaurant and began wandering around the night-club district, renewing our search for the pickup scene we had heard so much about in previous weeks. As Jane and I veered down a lane toward a pool hall, we stopped suddenly at the sound of some brash disco music raining down in droves. We looked at each other in surprise: Sano had no disco.

We followed our ears up a vacant stairwell that opened onto a vacant karate studio converted into a makeshift dance hall for the night. The narrow fluorescent lights had been replaced by frosty purple Day-Glo bulbs; portable strobe lights hung from the ceiling. A man dressed in a black leather jump suit, with a six-inch gold dollar sign strung around his neck, was performing a Japanese-language rendition of Grand Master Funk. I felt as if we had stepped through the looking glass into a SoHo under-ground nightclub.

Despite the elaborate decoration, we were the fifth, sixth, and, as it turned out, last persons to arrive all night at this ex-perimental disco. The proprietor, anxious for bodies to fill the vast open space, ushered us to seats on the wooden benches and gave us each a free cold Budweiser beer. No one was dancing, as all the guests sat huddled against one wall like nervous fourteen-year-olds, watching the scattered reflection of the strobe lights against the full-length karate mirrors and coughing every sev-eral minutes when the artificial fog machine inundated us with smoke.

The woman next to me sat stone-faced with a serene smile frozen on her lips and an Indian tapestry scarf around her shoulders. After several moments of awkward silence, she leaned over to show me how the black light glittered on her contact lenses, an effect that gave her eyes the eerie purple glow of a Siamese cat.

"Good evening," she whispered in a husky voice. "I met you before. Do you remember?"

I had no immediate recollection, but from the penetration of her eyes and the tone of her voice I could tell she was not a junior high school teacher.

"Oh sure," I said unconvincingly.

"I am the one who sells natural foods," she whispered. "We met on New Year's Eve."

Moko and her husband, Mino, she reminded me, owned and operated a small, natural foods restaurant in town called Rasa, or "Enlightenment." They served a small clientele of independent farmers and struggling artisans who lived scattered throughout the hills of Sano. "No married people come," she said, "and only a few men. Mostly women who want to start their own communes and make their own thread."

"What kind of food do you serve?" I asked.

"Whatever anyone wants," she assured me. "No meat, of course, and no chicken. Mainly brown rice and vegetables. Also no fish."

"You mean you don't eat sushi?"

"Oh, I love sushi," she said, pulling closer. "I eat it with my friends, but don't tell my husband. He doesn't know."

Finding it difficult to talk over the blare of the Sex Pistols and the glare of the lights, she and I wandered out into the foyer, leaving Mino to dance with Jane. Mino, she explained, was her second husband. For ten years she had been married to a salaryman in Tokyo, living a blithe domestic life with one child and a conventional husband who went off every morning to work in a construction company. For many Japanese this would represent the ideal middle-class lifestyle, but for Moko it was dull. "I couldn't do what I wanted," she told me. "I had little time to see the sunset." She divorced her husband and married Mino, a long-haired, bearded musician who had just returned from a five-year stay in India. Together they began the restaurant. "We have no money," she said of her new life, "and no time. But my heart is much cleaner. Now at least I feel better."

Her parents, she said, were sympathetic, but could not entirely understand why she would sacrifice stability in search of sunsets. "Intellectually they understand," she explained, "but really they just aren't involved. After all, they still eat white rice and meat."

We returned to the studio and joined the others for a dance to the rap music now on the stereo. The singer slid around the wooden floor in black-and-white-striped socks, banging two Budweiser cans together as if they were bongo drums. On the back of his leather jacket, two patches were sewn together. One of them said, in English, "WHEN I DIE I'M GOING TO HEAVEN, BECAUSE I'VE ALREADY SPENT MY TIME IN HELL — U.S. MARINE CORPS." The other showed a silhouette of two fighter planes and a battleship against a replica of the Japanese imperial flag. Its message: "REMEMBER PEARL HARBOR."

"By the way," Moko said, still twisting awkwardly to the music, "please don't think that I am not a good Japanese."

"Why not?"

"Because I think I *am* a good Japanese," she said. "I may not be normal, but at least I'm natural."

Jane and I said our good-byes and made our way out of the karate studio. At the bottom of the stairs I noticed a mimeographed announcement taped to the wall. "Everybody is invited to a new Saturday Night Disco," the bill declared in English. "Our name: The Weirdie's Party."

By now our search for the local singles scene had been deflected by the discovery of this thriving contingent of nonconformists living out their personal fantasies on the outer banks of the Japanese mainstream. Certain that we had happened on several flukes in the system, we ventured into a hostess bar, the type of place most often frequented by Japanese businessmen, and one that would surely have a more "normal" clientele.

The club, called Yōsei, or "Request," had a musty atmosphere with smoke lingering in the path of yellow spotlights, black velvet walls, and the kind of green velour couches that sag in the middle and give the customer no choice but to recline. The couches were half empty when we arrived at twelve-thirty, but before we had a chance to turn around, our coats disappeared into the hands of three Filipino girls, who smiled and welcomed us in both Japanese and English.

This type of establishment, known locally as a "Filipino bar," has emerged in recent years as the latest addition to the floating world of the Japanese "water trade." Enterprising club owners trek to the Philippines, peruse the bars and brothels, and convince young dancers and prostitutes to come to Japan for a chance to get rich quick. Because most of these girls can make more money in a single night of pouring drinks and giving massages in Japan than they can in a month in the Philippines, they readily agree. Over fifteen thousand such girls live in Japan at any one time. The Japanese government, aware of this new international geisha trade, grants each of these hostesses, often teenagers, a two-month working visa. The visa can be renewed twice, for a total stay of six months. Once in Japan, the girls are often crammed into small apartments and forbidden to have social contact with Japanese men. They learn a small vocabulary of necessary Japanese phrases, such as "Would you like to dance?" and "Do you think I'm beautiful?" as well as a few traditional songs to sing along with the guests. The men who flock to the clubs pay from seventy to one hundred dollars for two hours' entertainment above the waist, and twice that amount for an hour below the belt.

"Excuse me," said a girl named Candy as she downed a free snack of spinach and sesame paste, "can you use chopsticks?" Obviously she had been trained well.

Just as several of the girls nestled down on our couch, we were joined by a man in a lime-green blazer, a black silk shirt, and a large swath of hair that draped over his forehead. From appearance alone it was clear that he was no company hack. Hideo, as he told us, was thirty-one and ran his own antique shop and homemade jewelry store. Instead of offering the standard fare of Buddhist icons and lacquered cabinets, Hideo sold American army paraphernalia and cultural relics from the 1950s. "The fifties were great," he said. "It was the time of rockabilly, Elvis, and the New York Yankees. It was America's time."

"The Yankees?" I questioned.

"Sure, the Yankees were the ultimate," he enthused. "When Japanese people think about the 1950s, they think about the New York Yankees."

In addition to Yogi Berra and Mickey Mantle, other cultural idols from the 1950s such as Carl Perkins and James Dean blanket coffee shops and youth hangouts from one end of the country to the other. In Shinjuku, a popular youth section of Tokyo, one department store hung a hundred-foot banner of Marilyn Monroe from its façade during a promotion. A bestselling James Dean poster encourages lovelorn teenagers to think for themselves and preserve their individual dignity:

> *Self-deceiving love can be unsure.*
> *Keep a level head.*
> *Maintain your pride.*

Such symbols have become the slogans of a new generation of trendy Japanese, the *shinjinrui* I learned about during my weekend in the Snow Country. Hideo sold these "new types" all manner of "old-fashioned" American memorabilia.

Jane, for her part, saw more than a coincidental connection between contemporary Japan and 1950s America. Japan today, like America thirty years ago, enjoys sudden prosperity after a long spell of fiscal hardship, and many Japanese are starting to believe that their unprecedented national wealth entitles them to spend more money for their personal comfort and satisfaction. As a result, Japanese women have begun buying fancier home appliances like electric bread makers, men have begun turning to fancy European cars, and students have taken to eating imported ice cream, playing video games, and worshiping such hedonistic heroes as Elvis Presley and the Beatles.

Hideo looked to the American spirit as his model. "America was great in the 1950s," he said. "You had the Frontier Spirit." Although he spoke in Japanese, this expression, which constitutes the one pure, unadulterated fact that all Japanese students learn about the American character, came out in English. "America is very big, with a lot of land," he continued, "so Americans can live on the frontier. Japan, on the other hand, is very small. We have no land, so Japanese are boorish and unromantic."

Hideo chose to live according to the principles of happiness and self-indulgence, which he saw as the pearls of the American way. This goal led him to turn away from the well-trod path of cooperation in favor of an independent existence. He did not attend a university. He did not join a big corporation. He started his own store instead. "Japanese companies have lifetime employment," he said, "so the workers have no Frontier Spirit. It is the free merchants, like me, who have this feeling."

Not only did he personally reject the canon of Japanese thought which stresses "groupism" over individualism, but he felt that younger people in general were turning to this new philosophy, the Way of the Yankees. "Nowadays young men are becoming more individualistic, whereas in the old days they thought only of the group," he suggested. "Now, each man can discover new worlds."

As Hideo grew more and more animated, our Filipino

hostess grew more and more concerned that she was not fulfilling her duty to make us happy. She thrust a songbook into Hideo's lap and requested that he sing a song on the *kara-oke* stage. To press her point, she stood up and started to clap. Soon the whole bar was applauding in suit, and Hideo, unable to resist such temptation, took his turn at the sing-along stereo system. His song proclaimed his manifesto: to others' will I never kneel; I will go the way I feel. He sang the Frank Sinatra standard "Myyyy Wayyyy."

By the time he finished, half of the room was singing along as Hideo flipped his hair back and soaked in the spotlight. When he returned to our table, he made one final observation. "Japanese people are all yellow monkeys," he said, repeating a common self-effacing slur. "We are a different color from you. Nevertheless, the Frontier Spirit does not care about color. Even the Japanese can have it."

Finished with his speech, he picked up a napkin, wrote something down, and thrust it into my pocket. Then he asked Jane to view a "late-night movie" with him, and the two of them left the bar. For the first time all evening I was left alone with the bill — $120 — and I set off quietly for home, the lonely victim of the pickup scene I had set out to uncover.

I wandered the empty streets of Sano toward my apartment, past the barbershop, the billiard hall, and the "Swan Song" music store, which had closed down the previous year. Each of these buildings stood apart from the others, but the mist of the night and the pale moonlight seemed to draw them together in a single line. As I walked, I thought of the people we had met that night. While others view Japanese society as strict and unforgiving, these individuals believed it to be resilient enough to incorporate them. Although they rejected traditional definitions of Japan, they did not flee its boundaries but broadened them instead. Even alone, in the middle of the night, they kept in sight of the Sun.

As I stood in the cold outside my door, fumbling for my

key, the napkin from Yōsei fell out of my pocket. I opened it to read the message Hideo had written. "Japanese people are one race," he had scribbled, "so we always return to the land of our ancestors. We love the cultures of all other worlds, but most of all we love Japan." At the bottom of the page he had written his name, the date, the hour — 2:00 A.M. — and a simple message in English: "Good Life. I love you."

17

THE WAY OF LOVE: HOW TO PICK UP A JAPANESE GIRL

If anyone among this people knows not the art of loving, let him read my poem, and having read be skilled in love.

— *Ovid*, The Art of Love (*I:1*)

"YOU HAVE A DISTINCT ADVANTAGE," the bartender shouted from behind his counter. "Japanese girls love blonds."

During my search for the secrets of *nanpa*, I often heard this remark. Few people claimed to know much about picking up girls; even fewer confessed to having done it; but everyone agreed that I would find it easy because I am naturally blond. Actually, my hair is brown, but that didn't seem to matter: the true secret to *nanpa* lay in marketing.

One rainy Sunday night in early March I got much closer to the heart of *nanpa* when I found myself in a quiet one-room coffee shop in Sano with the unlikely name "Potato," in the hands of two self-proclaimed experts who had volunteered to teach me a few tricks of the trade. From the pickup line to the check-in desk, they included every step.

"You have to think ahead," began Ishikawa, the owner of the SPEL *juku* who had brought me to this coffee shop to meet his friend Sato, the master. "You have to go to a place where the girls are. Remember, the early bird catches the worm."

Sato, who like his friend was married with one child, confirmed this strategy. He recommended a department store, a girl's natural lair. There the cunning fellow scouts the ladies' floors, spots an appealing prospect in the linen — or kitchen, or bathroom — department, and approaches with this line: "Excuse me, I was just trying to buy a pillow — or toaster, or towel — and I was having a bit of difficulty making a decision. I wonder if you might be able to give me some advice?" This strategy is advantageous, they agreed, because it creates an artificial debt which the gentleman owes the lady and which he insists on repaying immediately — say, at the coffee shop on the top floor of the store.

While this approach has been time-tested and is virtually geek-proof, Ishikawa insisted that the absolutely best place to meet a girl is on a ski slope. There, the man, in his natural capacity as "master," offers to teach the girl, in her intrinsic state as "disciple," the most up-to-date alpine techniques. This approach is preferable, my *sensei* said, because it rekindles memories of school and the seminal teacher-pupil bond.

This initial meeting qualifies as the first date. The skillful man, however, must not conclude this introductory encounter without first acquiring his potential beloved's vital statistics: her name and telephone number. From here, he is ready for the second date.

"Sports are best," Sato advised. "When you play together, she develops an appetite. First you play, then you eat, then you 'play' again." Athletics, it seems, are the aphrodisiac of choice.

Short of sports, the *séducteur* could suggest some other leisure-time activity such as zoo-going or movie-watching. But the man's objective throughout this encounter should be to indicate his unflagging interest in the girl by making physical contact.

"Be sure and touch her," Sato stressed. "Of course the shoulders are best."

"But don't forget," warned Ishikawa, clearly the more prudent of the two, "you should praise her as much as possible. Hair is best, or the face. Just praise what you like about her."

"What if I think she is smart, or funny," I asked, "should I praise her mind?"

"Of course that's all right," Sato said, "but you better stick to her hair or ears or chin. Looks are more important than brains, you know."

Having arrived at the crucial third date, Sato and Ishikawa warmed to their subject and adopted a sincere, almost solemn tone. Sato decided to close down his shop for the night in order to devote his undivided attention to guiding me through the delicate stages of advanced courtship.

The third date, they announced, is the right time for the first kiss. After taking her out for the evening and treating her to a few drinks and a fine meal, the man must be ready to make his move.

"All it takes is guts," Sato said. "You gotta push."

"Walk her to the door," Ishikawa suggested. "Stand still for a moment. Then kiss her —"

"But only a light kiss," Sato interrupted. "Not too hard, not too deep. Give her a chance to respond."

"Then," his friend continued, "if she doesn't resist, you can take her in your arms. Just as I tell my students, slow and steady wins the race."

By this time, if the prudent man has proceeded with caution through the previous stages, lavished praise on his lover's locks and whetted her thirst, he is ready to advance to the final frontier of romance in Japan: the "love hotel." These modern mosques of passion, which the Japanese call *labu hotelus*, are as good a symbol as any for contemporary Japanese culture. They mark a unique confluence of architectural ingenuity, lasciviousness, cuteness, and efficiency — attributes that would appear high on any list of Japanese character traits. An estimated twenty-five thousand of these unabashed, neon-blazoned dens of desire fill the downtown alleys of major cities and the outer reaches of rural towns. Constructed to look like medieval castles with towering turrets or cruise ships with soaring bow and stern, love hotels rent rooms by the hour (a "rest") or for the night (a "stay") and provide "easy in/easy out" service. They are, in short, the fast-food franchises of love.

"It's really very easy," Ishikawa observed. "No problems at all."

The rooms in a love hotel come in assorted shapes and sizes, with decorations based on taste. They offer everything from beds in the backseats of Benzes to nests atop artificial golf greens, from marble Jacuzzis in the bathroom to "sense-a-round" body massage in the mattress. Many rooms are equipped with an extensive bank of stereo, video, and *kara-oke* equipment, as well as an audiotape library that contains everything from country-and-western ditties to Zen Buddhist mantras. These music boxes also have a special selection labeled "ALIBI." The alibis are tape recordings of sounds from public places such as a street corner or bowling alley and can be used as background music for a hurried telephone call back home: "Sorry, honey, I'm afraid I'll be a bit late tonight. Stopped off at the club for a round of golf with the guys." Like any good service industry, love hotels also provide a guest book in each room, complete with spaces for lovers to inscribe pet names for themselves, describe where they went on their date, and circle any of two dozen diagrams that depict various lovemaking positions. Finally, for convenience, a condom is included in the price of the room and is located, Sato noted, "just under the pillow."

To help an anxious couple make the proper room selection, photographic slides depicting each room are displayed on a lighted board at the check-in counter. For choosy lovers, this board serves the same function as the flavor chart inside a box of Whitman's chocolates. This way for creamy nougat; that way for cherry bonbons.

"Choose a room with good feeling," Sato recommended. As an example, he noted that Japanese girls around twenty years old like Disneyland, so perhaps a room with Mickey Mouse sheets would help the cause. "Atmosphere makes her want to do it," he observed.

Having led me this far, these two men were not about to leave me stranded. After offering me a plate of potato chips and a glass of peach-flavored milk, they began to advise me on the proper way to make love to a Japanese girl.

Begin with the skin, they advised. It seemed somehow fitting that sex, like so much else in Japan, would begin with attention to surface detail. "Japanese skin is very soft," Ishikawa explained with a misty gleam in his eyes, "just like Japanese *mochi* rice cakes."

Starting with the head, the gentle lover should proceed downward, he said, paying special attention to the ears, the neck, and the breasts. Arriving at the waist, though, the wise one skips to the toes and eases up the smooth, sheer legs to that special spot, "like Gold Finger, with a soft, soft touch."

As these two married men with young babies leaned over the bar, leering at their hands as they traced luscious images in the air, suddenly their eyes lost their sparkle and their arms went limp. They dropped their hands to the counter and looked first at each other and then at me.

"Oh no," Sato wailed. "You can't do it . . . You won't fit!"

"What are you saying?" I protested.

"Japanese girls are too small," Ishikawa declared, "especially in that *important* place. For Japanese men it's okay, but for foreigners . . ."

Was I, in biological fact, being denied access to this secret sanctum of Japan, or was this just another artificial trade barrier? After all, if I could use chopsticks, then surely I could handle this challenge.

But Sato and Ishikawa were deadly serious. Sato reached across the counter and grabbed a bottle of Tabasco sauce.

"You see this?" he said, waving the bottle in front of my

face. "This is your average Japanese man — about ten to thirteen centimeters long. You are longer, right?"

"Uh, well . . ."

"It's true," he burst in. "But although yours may be longer, mine is harder. Japanese men are short but strong. American men are soft."

While I pondered this new cross-cultural theory, Sato disappeared briefly into the kitchen and emerged brandishing an empty teakettle that he proceeded to fill with water. With great flourish and bravado, he held the Tabasco bottle to his crotch and suspended the teakettle from the neck of the bottle. *Ta-dum.*

"There," he shouted, as if he had just pulled a rabbit from a hat, "I can do that . . . Can you?"

"I'm afraid I don't know," I replied humbly, hoping he would resist the temptation to test my virility right there in the middle of the Potato. "I never tried. But does this really mean I can't go to a love hotel?"

"Let's just say," Ishikawa remarked, "that it would be difficult."

Just then Sato had an idea. "You know, Japanese girls can give birth," he said, by now grossly inflating the size of foreign men. "Maybe you *can* do it. You just have to be careful and make her relaxed."

"That's why atmosphere is so important," Ishikawa added, his hopes inflating again. "You must have a good room, with mirrors on the wall, a round bed, and a see-through shower. That will help."

"Of course, don't talk to her," Sato added as an afterthought. "Keeping silent is not so difficult, I think, and it's a lot more fun. Just think of it as a kind of exam."

For a moment the two men disappeared into their own memories, flashing faint smiles of recollection and nodding intently. Ishikawa was the first to break the silence. He leaned forward, slapped his friend on the shoulder, and said, "You know, we should know better at our age."

"We should," his friend conceded, "but . . ."

"You," Ishikawa said, turning back toward me. "For *you* to know Japanese girls is important."

"Indeed," Sato agreed. "By getting to know many girls, you will become a good man. You will have a warm heart, a strong body, and eventually . . . a better mind."

It was comforting to learn upon arriving at the end of this odyssey that all this plotting and maneuvering would lead not only to carnal knowledge but also to the loftier end of personal enrichment. The inexperienced outsider could only wonder if one could learn the *wabi-sabi* — peaceful enlightenment — by traveling down *this* road.

But short of that, Sato had one more piece of advice before sending me out on my own into the jungles of Japanese dating. "Don't worry if you don't succeed at once," he said with a conspiratorial wink. "If you fail, you can always use your hand. When I was a young man, my lover was my right hand."

18

POMP AND CONSEQUENCE: GRADUATION DAY

May thy peaceful reign last long:
Beyond ten thousand years.
Until what are pebbles now,
Into mighty rocks shall grow,
Which graceful moss doth line.
— *"Kimi Ga Yo," the Japanese national anthem*

ON THE DAY BEFORE GRADUATION, during a blustery mid-March storm, all the students of Sano Junior High assembled outside the gymnasium in a seemingly endless single-file line. At the sound of a shot from a starter's pistol, the students began a silent procession through the double doors of the gym. As the brass band rehearsed patriotic music under the basketball hoop and teachers draped red and white bunting from the walls, the students made their way down the center aisle, pivoted left or right per instruction, and assumed their assigned seats. When all the classes were finally settled, the principal took to the stage.

"*Kiritsu*," he shouted, and the students rose to their feet.

"*Rei*," he said, and they bowed in silent submission.

"Tomorrow is graduation day," Sakamoto-*sensei* boomed into the microphone, "the most important day in the year. Tomorrow our school must shine."

The average Japanese student will graduate four times in the course of his or her education — from elementary school through university — but none of those graduations is more important than the junior high school ceremony, which marks the end of compulsory schooling. Until this day, students have attended the school closest to their homes, but with the onset of high school, academic performance becomes more important, competition increases, and neighborhood groups are forced to dissolve in the face of society's selection process. The ninth graders at Sano Junior High, having recently passed their entrance exams, would soon be dispersing to ten different high schools, ranging from the topnotch all-male school in neighboring Ashikaga to the agricultural trade school a one-hour train ride away in the hills of central Tochigi. With this separation looming just over the horizon, graduation day from junior high formally closes the era in which the school has served as the unconditional haven and protector of all.

"The way to shine," the principal admonished his students, "is to work as one school. If one of us is out of line, all of us look bad. Listen closely to your teachers; watch your classmates carefully; and stand together as one. Today we will polish, so tomorrow we can shine."

With so much importance focused on the ceremony itself, schools go out of their way to ensure that their legions are fully prepared. Although students at Sano had been bowing together for close to ten years — in every class, at every assembly, before and after every lesson — they would spend this afternoon reviewing for one final time the basic marks of social punctuation: how to sit, how to stand, and how to bow.

"First, put your legs together and rest your arms on top of your knees," the principal instructed. "Now move to the end of your seat . . .

"Don't wiggle in your chairs. Remain still until you are called to stand . . .

"When you bow, remember to count to yourself — one, two, three — then lift your head and look toward the front of the room . . ."

Learning to bow is a continuous process that students must practice from their first day in school — along the highway — to their last — in the gymnasium. Bowing, derived from the Confucian custom of showing trust toward strangers by dipping the head in respect, is far more than a mere social greeting: it is the prime ingredient in the syntax of Japanese life. Students learn that by adjusting the depth of their bows, they can express themselves as effectively with their bodies as they can with words. In this vernacular, a friend rates only a slight nod of the head, while the principal merits a full-body bend in which the hands rest lightly on the knees and the head dips down below the waist.

"You seventh graders are not together," a teacher called from the catwalk above.

"Hirohashi-*san*, don't smile when you bow to the front."

"Look straight ahead at all times."

For two hours the students repeated these maneuvers, working methodically, retracing steps, honing their execution. Although the practice seemed endless, the legions did appear to be making progress, and by the end of the afternoon the entire student body was able to stand and sit without flaw. Upon command, they would rise in silent union, bend with the supple ease of swaying wheat, and return to their seats in effortless grace like a spent wave sliding back toward the sea.

"If you can bow like this tomorrow," the principal declared when the practice was over, "your parents will be proud, your teachers will be proud, and you can be proud of your school."

At eight A.M. the following day, hundreds of adoring parents crowded around the double doors of the gymnasium, all vying to be the first to fill the folding chairs that had been laid in even rows across the recently revarnished basketball court. Before entering the hall this Thursday morning, they tucked their shoes into tiny plastic bags and donned slippers brought from home. The sight of over six hundred mothers in silk kimonos and a half-dozen fathers in black tuxedos sitting expressionless in tiny metal chairs, with their feet inside pink and yellow slippers dangling just above the floor, will always be one of my fondest memories of Japanese schools. In addition to the parents, special representatives from the mayor's office, local businesses, and the PTA all joined with the brass band at the start of the ceremony to hum the year's theme song: the Beatles' "Let It Be."

After the students had marched down the center aisle and bowed in unison toward the empty stage, the first person to address the gathering was Mogi-*sensei* from the Board of Education, the teacher who had sent me tumbling into the hospital six months before with his lethal forehand dropshot. He walked to the center of the cavernous stage, stopped before the free-standing microphone, and began to read from a handwritten scroll.

"I come today to the most outstanding junior high school in our community to extend to the honorable parents, the hard-working teachers, and of course the bright and cheerful students, my sincerest congratulations on the occasion of your graduation . . ."

The speech continued like this for almost ten minutes as Mogi-*sensei* recounted the educational odyssey that the students had completed, extolled the virtues of hard work that the school had instilled in them, and waxed poetic about the outstanding citizens they would soon become in selfless devotion to their country. At the conclusion of his speech Mogi-*sensei* offered a deep and gracious bow, tucked the scroll into his jacket, and marched quietly down the stairs. The students nodded their heads in respect. The teachers beamed with pride. Several mothers began to cry. Unbeknownst to them, Mogi-*sensei* would deliver this same address at no fewer than twenty-five schools during the coming two weeks.

Following this salutation, the chairman of the PTA, the assistant principal, and the chief of the Fire Department each marched to the identical spot on the stage and delivered almost identical speeches. The audience seemed transfixed. From the front row of ninth-grade students in their black uniforms to the last row of parents in their black and white kimonos, no one moved throughout these proceedings: the mothers never smiled; the students never laughed; the audience never applauded. With the conspicuous silence, the mourners' clothing, and the grim demeanor, the ceremony felt more like a requiem to lost youth than a celebration of life to come.

To add to the solemnity, each person who marched across the stage, including the graduates who went one by one to receive their diplomas, partook in an elaborate ritual: students bowed to teachers; teachers bowed to guests; guests bowed to parents; and everyone — absolutely everyone — bowed to a five-by-seven-foot Japanese flag, which stared at the congregants from the back of the stage like a Cyclops come to life. After remaining invisible for most of the year, the "circle of the sun," or *hinomaru* flag, suddenly emerged in all its glory on graduation day.

A national flag is a relatively new invention for Japan. Banners showing the sun were used by some samurai clans, but the current design was not adopted until 1870, after Commodore Perry arrived from the United States and forced Japan into the international community. Almost since its adoption, however, the flag has aroused controversy. Across Southeast Asia the *hinomaru* flag quickly became the symbol for Japanese aggression. For centuries Japan was isolated and did not need a national emblem, but less than fifty years after hoisting its first banner, the country began terrorizing its neighbors. Many Japanese feel as negative about their flag as other Asians do. Since Tokyo's defeat in the war, both the flag and the national anthem — a dreamy tribute to the emperor adopted around the same time as the flag — have been an undying source of bitter memories for some citizens, especially those on the southern island of Okinawa who suffered the country's only direct invasion by Allied troops. As a result, the striking red sun on a plain white field, which the rest of the world has come to associate with Japan, has been virtually locked out of sight in the nation's schools for most of the last forty years.

Recently this mood has shifted, and the flag and the anthem have been staging somewhat of a government-sponsored comeback. While protests and flag burnings still occasionally occur, new regulations require schools to raise the *hinomaru* flag and play the Kimi Ga Yo anthem at all school ceremonies. Critics complain that the policy will prompt a return to Japan's military past, but the government claims that learning to respect the Japanese flag is the first step in respecting the flags of other nations. In effect, the government is only making explicit in schools what has been implicit all along — that students must learn to bow not only to their teachers but to the state as well.

After the preliminary speeches and the distribution of diplomas, the final person to address the students was the principal himself. Standing erect before the flaming sun, his black suit sharply creased, his black tie clipped to his starched white shirt, his gray hair slicked back like polished grain in marble, Sakamoto-*sensei* seemed to embody all the power and dignity of the state. For a moment he surveyed the scene below him: the long rows of students like beads of oil on water, each one glistening independently, but also melding with the others in a common glossy bond; the parents seated quietly behind their children, flushed with pride at the solemn charge being given their sons and daughters. For the school, as for the principal himself, this was the high mark of the year. He was "one-man" in control.

"Winter is over," he said in a booming voice, waiting for his echo to bounce off the walls before he spoke again. "Spring is coming. Soon the cherry blossoms will brighten our neighborhood and remind us how fragile we are. This is the season of change . . ."

In the days leading up to graduation, I had learned how obdurate the state can be on the subject of change. About a week before the ceremony Denver asked me if I would like to deliver a short speech to the school. Several students had suggested the idea, he said, and many of the teachers had agreed that it would be a wonderful opportunity to add an "international" perspective to the occasion. I began preparing what I would say. But two nights before graduation Denver came to my apartment long after midnight and told me I would give no talk.

"The principal said it is not appropriate," Denver explained, his voice betraying his distress. "Graduation is a Japanese ceremony, he said, and the format cannot be changed. Since you are a foreigner, he was afraid you might upset our customs. We cannot compromise tradition, he said."

Sitting in the audience listening to the principal address the school, I felt the weight of that tradition — the burden of a man who was schooled in the past, and who, when faced with the option of the future, chose the comfort of the old over the risk of the new.

"Each of you has worked hard since your days in elementary school, " Sakamoto-*sensei* said to the students. "Now you have the skills necessary to become a successful *shakai-jin*, a member of society. But I must tell you that your training in life begins now. Graduation marks a change in your life, like moving from one room to the next. But please remember that you must never leave an open door behind. At home, when you go from one room to the next, you always close the door behind you. The same happens when you leave this school today. You must draw a *kejime* with your former life."

Kejime, which can be translated roughly as a "line of demarcation," is one of the primary pillars of Japanese education. Originally the name for a bamboo screen that divided two rooms, *kejime* today implies a code of behavior that one accepts upon reaching a certain status in life. High school students earn more free time than junior high students but have the added burden of preparing for university entrance exams. Workers enjoy the perquisites of company life but must accept long working hours in service to the corporate good. The principal may exercise strict control of his school but in return must shoulder the responsibility if any of his troops step out of line. *Kejime*, in essence, is the law of the land: as one grows from a pebble into a stone, one gives up certain personal freedoms in return for the benefits of achieving a higher position in the community.

"As you leave here, you will all strive hard in your new lives," Sakamoto-*sensei* declared at the end of his speech. "But please remember what you learned in this school: work comes before play, the past before the future, and the snow before the spring. Your parents, your teachers, and your country are depending on you. We hope you do your best."

At the conclusion of the ceremony, all the seventh- and eighth-grade students lined up outside the gymnasium in the form of a giant human funnel, like cheerleaders before an American high school football game. As the graduates marched down this open aisle, small groups of underclassmen rushed forward to say goodbye with bouquets of flowers and folded farewell notes. After the processional, the graduates gathered just outside the school's main gate to savor the moment, hugging, crying, and snapping photographs of one another. Many of the younger girls asked their male idols for buttons from their uniforms to keep as mementos. The girls provided the scissors, and the boys from the graduating class spent the next thirty minutes carefully snipping buttons from their Prussian army-style jackets and presenting them to packs of shrieking girls. Some of the boys were so popular that they soon finished off the buttons on their black jackets and had to begin clipping the ones from their plain white shirts. Even in the excitement, however, the boys were careful never to remove the second button from the top of either their jackets or their shirts, for to hand over this token to a girl meant much more than platonic love and implied some sort of "commitment."

After mingling at the gate for nearly two hours, the graduating seniors finally dispersed for more private celebrations away from school. A group of about fifteen students moved to a nearby community center to have a small party of their own. Here, though they were far from school, trouble began.

Based on what I later learned, this party was innocent enough: there were no drugs, no liquor, no vandalism, and no sex, just a coed group of fifteen-year-olds sipping sodas, eating chips, and listening to Madonna on a cassette tape player. The students had even reserved the room in advance. The problem was that the students held their party at two-thirty in the afternoon, during official school hours. But hadn't they just graduated?

"The students are the responsibility of the school," the principal explained to me after he had summoned the students back to school, reprimanded them, and sent them home with a warning not to indulge in such debauchery again. "Even though they graduated today, they remain our responsibility until the end of the month. Then they move into the care of their high schools."

"What about their parents?"

"They may be at work," he said, "or out of the house. Or maybe they just don't care."

Such is the essence of child welfare in Japan. As students cannot be entrusted to care for themselves and cannot be left to the random and perhaps less-than-strict rule of their parents, they fall under the complete dominion of the schools. Until the students are finally turned over to someone else the state can trust, the school must open and close the doors of *kejime* for them.

"After students leave our doors," the principal said, "they essentially break from the school. They no longer *belong* to us. If they return, they must stand at the door of the teachers' room and announce their names and their reason for entering, just like any other guest. The school cares for students until the end of March. Then we let them go."

Considering the severity of this transition, it was no wonder that the ceremony itself proved to be so somber. Graduation is only one in a series of benchmarks that Japanese experience as they are handed from one group to another, all the way through their lives. Given this legacy of supervision, the sense of *amae*, or dependence, that many Japanese feel toward their employers and their colleagues seems easier to understand: most have never been out on their own.

On the last working day in March, all the members of the prefectural government who had offices in Sano were called together in one room on the top floor of our office building. Dr. Endo and his Health Department staff were there, as were the forestry, housing, and finance bureaus. Each department stood in a double-file line. Arai-*san* and Eh-*chan* wore flowered kimonos and stood in the back of the space allotted for the Board of Education. At the front of the room, on a tiny stage, three objects were lined up in a row: a flag of Japan, a flag of Tochigi, and a Sony reel-to-reel tape recorder on top of a metal folding chair. When the meeting was called to order, an attendant reached over and turned the recorder to "PLAY."

"Good morning, ladies and gentlemen," said a scratchy male voice on the tape. "Thank you for coming today. The governor wishes he could be with you this morning, but instead he will now deliver his year-end greeting.

"*Kiritsu*," the voice shouted, and everyone snapped to attention.

"*Rei*."

And without further instruction, the several hundred professionals who had assembled in the fifth-floor meeting room of a white stucco building on the outskirts of Sano-*shi*, including my boss, my doctor, and even myself, bowed together to a tape recorder that sat on a chair in the front of the room. At that moment, as my head dipped toward the floor and my palms brushed over my knees, I realized that Japanese students can never claim that what they learn in school will not serve them in the "real world." In Japan, it most certainly will.

19

COUNTRY BUDS AND FADED DREAMS: A CHERRY BLOSSOM SPRING

The spring has come, and once again
The sun shines in the sky;
So gently smile the heavens, that
It almost makes me cry,
When blossoms droop and die.

— *Kino Tomonori, c. 905*

CHO CAME TO MY HOUSE EARLY. Tufts of clouds drifted across the azure Sunday sky, and the pale green blush of newborn leaves enlivened the drowsy air. Instead of heading far away on this morning, Cho took me back to his hometown for the annual April extravaganza known as the Kuzu *Genjin* Cherry Blossom Jubilee. This celebration draws people from all across the Kanto Plain into the quiet mountain hamlet of Kuzu to celebrate the coming-out of the national flower, *sakura*, and to honor the town's most famous son: the Kuzu *Genjin*, the oldest human remains ever found in Japan. "He's in every high school text-book," Cho boasted like a proud father. Although the esteemed *Genjin-san* amounts to a mere three broken bones and a sliver of skull, he is immortalized on T-shirts, banners, and coffee mugs as a squat, affable man with a bushy beard, an animal-skin tunic, and a bulging hunting club. "Discover Kuzu *Genjin*," the town slogan cheers, "Oldest Man in Japan."

After a fifteen-minute drive north from Sano over two-lane highways pocked by the constant trampling of dump trucks and cement mixers, Cho parked his Nissan Sunny behind the elementary school where his father served as principal. The three-block area of downtown Kuzu had been gussied up for the fair: pink streamers wafted from every light pole; red and white tents lined the side of the main street, where vendors hawked hard-boiled eggs soaked in soy sauce and bananas draped in chocolate; a green portable house of mirrors had been wheeled in front of the town hall.

"*Sensei, sensei,*" a girl called to Cho, "do you want to go inside?"

"No, thank you," he answered. "I'm scared of ghosts."

As we weaved through the maze of food stalls and up the daffodil-lined stairs toward the cherry grove at the top of a hill, Cho began to relate the sad story of Kuzu. Like a boom town in the American West, it had sprung from nowhere a century ago after a rich lode of natural stone and chalk was uncovered deep in the hills just outside of town. For several generations the stone from these quarries was used to cover buildings and build bridges all across the archipelago.

"Thirty years ago, Kuzu was a rich place," Cho explained. "We had many big companies, and the big companies paid big taxes. We were a developed town."

But over time the big companies merged with bigger companies and developed even bigger interests beyond the mines of Tochigi. Eventually most of them moved their headquarters to Tokyo. In a story that has been played out in rural communities across time, the hinterlands — rich in natural resources — became subjugated to the city as managers took the profits with them and left the workers to labor in the country, slowly stripping themselves of their only source of wealth. With most of the mines soon emptied of their cache, stagnation slowly spread across Kuzu, then malaise, and finally the inevitable disintegration.

"Now that all our companies are gone," Cho said, "all we have left is dust. We used to produce the finest stone in all of Japan, but now we make only chalk that is used to line high school baseball diamonds." He stopped on the stairs and scanned the decaying tableau of tin-roofed buildings, blistered roads, and barren mines — like open sores — that scarred the surroundings of the somnolent town. "When I was a student, almost thirty thousand people lived in Kuzu," he said. "Now I'm a teacher, and that number is less than half."

We arrived at the top of the hillside park, beneath a thicket of arching trees that erupted into a gossamer of miniature blossoms. The pale pink branches reached across the sky and linked arms overhead in a cotton-candy arcade. Children climbed up the boughs and out on the limbs to shake the petals from their clusters into a shimmering pastel rain. Their parents sat on straw mats on the ground, drinking beer from plastic kegs, roasting beef on open grills, and belting *kara-oke* classics from battery-generated portable stereos: "It's spring; the cherry blossoms are warm; my heart blooms for you . . ."

The first time I heard about *sakura*, I assumed that after the deluge of blossoms in early April, most of the country would be rolling in plump, juicy cherries. Not only was I incorrect — these trees bear no fruit — but even to consider such a thought was to miss the point of the season. *Sakura* is the national flower for a reason: the blooms come and go, tantalize and evaporate in a single moment of brilliance that transcends time. For centuries the *sakura* has symbolized a pure brand of beauty, tinged with elegy. During the Pacific War the cherry blossom became the motivating symbol for a special band of pilots who fought with sublime inspiration but knew that they would die. These martyrs were named *kamikaze*, "the Divine Wind":

> *You and I, companion cherry blossoms,*
> *Flowered in the garden of the same military school.*
> *Just as the blossoms calmly scatter,*
> *We too are ready to fall for our country.*

·   ·   ·

Autumn is a time to venture into the forest and seek tranquillity; spring is a time to wander among the trees and consider life's impermanence.

"In America we have an expression," I said after we stopped to buy a snack of tangy rice balls and green tea. "In spring, a young man's fancy turns to thoughts of love."

"I've been thinking about love," Cho said, not changing the solemn look on his face, "and I've made a decision."

"About what?"

"About Chieko . . . I'm not going to marry her. I've decided she's not the right girl for me."

I stopped for a second, but Cho kept moving as if nothing had happened.

"I'm sorry to hear that," I said, trying to catch up. "When did you decide?"

"About a month ago."

"Have you told anybody?"

"Only her. Otherwise, I have nobody to tell. But this is a small town. Soon everyone will know."

For the long months that Cho and Chieko were dating, they were constantly shadowed by the web of unwritten rules which still controls small-town life. During their courtship they never appeared alone in public. If they planned to have lunch on Sunday, they would drive to a neighboring city. If they wanted to see a movie on Saturday night, they would travel to the next prefecture. The contrast with life in Tokyo could not have been more pronounced. While Cho and Chieko were driving surreptitiously around south Tochigi lest a colleague spot them together, Hara and his girlfriend were parading around steamy nightclubs in Tokyo and returning to his apartment for casual sex.

"Life is different for teachers," Cho told me. "We are public servants. But what is bad for teachers all over Japan is even worse for us in the country."

Strolling among the cherry blossoms, I felt for one of the first times a gap between our lives. Did Cho not mention his breakup to anyone because nobody cared, or because he preferred to keep his feelings to himself? Did he want to talk about what happened, or let it pass silently? I resisted asking too many questions at first, for fear of intruding. I had often noticed the lack of public discussion in Japan about private lives. While the men in my office loved to go drinking, joke about sex, and make passes at *kara-oke mamas*, they rarely talked about their families or their lives at home. Japan does not have the ubiquitous call-in radio talk shows where people can confess their marital infidelities or sibling jealousies. I sometimes thought that the most telling difference between the United States and Japan might be that Americans thrive on this kind of public therapy and Japanese do not.

After several minutes of silence Cho did begin to talk. The story he told was less about two people who could not love each other than about two worlds that could not converge.

"Chieko and I are both teachers," he said as we moved to watch the finish of the annual *Genjin* Run. "She grew up in this area, and so did I. She went to high school here, and so did I. Then we both went off to university. But I went south to Tokyo, and she went north to Utsunomiya."

Utsunomiya, the capital of Tochigi, is a low-brow, low-tech city of several hundred thousand. Utsunomiya University, or Udai for short, was originally a teachers' college before the Second World War but became a "general education" school during the postwar reform movement. This reform, like many others, did not take hold, and today the university still produces an abundance of teachers, who in turn go on to populate area schools. Since most teachers in Tochigi graduate from this program, the prefectural school system is run like an exclusive club whose members work with their friends, marry their colleagues, and rarely leave the nest. In my office at the Board of Education, for example, fourteen of the sixteen teachers graduated from Udai.

Cho, because he went to a university in Tokyo, was excluded from this clique. "People I didn't even know would come up to me and say, 'I hear you are going to marry Chieko,'" he said. "I would wonder where they had heard this information, and then the person would say, 'I was her classmate at Udai.' No matter how hard I try, I can never get away from this circle. I will always be an outsider, just because I studied in Tokyo."

Cho also discovered that Chieko and he had learned different things while they were away. "I like to travel," he said, "but she does not. I like to think about the world beyond Tochigi, but she does not. She is content never to leave her parents' home. Everybody says that our students have to learn to live in a world that is larger than just Japan. But first, our teachers have to learn that this world exists."

Eventually, perhaps inevitably, Cho and Chieko parted ways. What seemed in the fall like a promising match, by the spring had withered and died. We walked among the falling petals alone.

"How do you like the festival?" Cho's mother asked me several minutes later when she found us waiting by the finish line. A former teacher herself, she was a pleasant woman with graying hair around her temples, round glasses clipped to a silver chain around her neck, and, on this special day, an indulgent touch of lipstick on her smile.

"Kuzu is beautiful in the spring," I said.

"This is our second most important festival," she said with the bubbling enthusiasm of a practiced tour guide. "The first is the fireworks extravaganza in the summer. I hope you can join us for that. This year's theme has just been decided: 'Come Back to Kuzu: Oldest Town in Japan.'"

"It sounds similar to the Cherry Blossom Festival," I suggested.

"We are trying to attract people from Tokyo this year. We want to increase our tourism."

"Cherry blossoms and fireworks are all we have left," Cho

said. "Maybe they won't notice that all our shops are closed."

"Hush now," his mother said, then turning toward me she added, "He's been very negative lately. I can't understand why."

"I think we had better be leaving," he said.

As we made our way down the hill, Cho began talking. "You know what they're going to do this summer? They are going to build a large replica of the Kuzu *Genjin* made entirely of fireworks. I can't believe it. It's going to be ridiculous."

He was silent for a second, then started to speak again.

"I want you to know that this isn't just Kuzu's problem, it's Tochigi's problem as well. Many people say that rural prefectures like this one will become more powerful in the future. 'The Age of Tochigi' will come in ten years, these people say. I'm afraid I don't agree."

For the first time in the many months that I had known him, Cho appeared to be getting angry.

"We have fast trains and fast highways in Tochigi, but we don't have any money. We especially don't have the spirit. Recently the head of a big company in Utsunomiya said that Tochigi was a backward place. 'The people are too easily satisfied,' he said, 'and until Tochigi gets a university that will produce workers with the right spirit, companies will not succeed.' I agree. Everyone with any sense is moving to Tokyo and taking our future with them . . .

"Look at Hara-*kun*," he said. "He was from the countryside and he moved to Tokyo. Now he has a wonderful job, a high salary, even a beautiful fiancée. In Tochigi we have nothing. Some men on farms have to import girls from the Philippines, just to find someone to marry. It's embarrassing."

"But Hara complains that Tokyo is too crowded," I said. "He can't buy a house because they're too expensive, and he can't even afford to look at a piece of land within two hours of his job. His salary does no good because everything costs too much. He wants to move away."

"Well, he shouldn't come here," Cho said, stopping at the

door to his car and fumbling with the key. He finally managed to unlock his door, then tossed his keys to me. "Don't get me wrong," he said, "I like Tochigi. I just don't like Tochigi people." A sly grin flickered on his face. "When I'm governor, I think I'll get a new population."

Tochigi falls just outside the golden crescent of prosperity that stretches south from Tokyo, through the regional centers of Nagoya and Osaka, to the southwestern port city of Kobe. This stretch of land, once known as the Tokaido Road, remains the most densely populated and developed region in the entire country, similar to the Boston-Washington corridor in the United States. As this zone has sucked in more and more investment over the last thirty years, many rural areas have stagnated because of their inability to satisfy the upwardly mobile ambitions of new generations.

Even schoolchildren know that life in the *tokai*, or big city, is different from their routine in the *inaka*, or countryside. Several junior high school girls in a nearby town approached my friend Jane one day while she was teaching at their school.

"Miss Jane," they asked, "have you ever kissed a boy?"

"Yes, I have," she replied, "have you?"

The girls blushed and covered their mouths in shock.

"Of course not," one squealed.

"I don't even have a boyfriend," said another.

"I'm only going to kiss the boy that I decide to marry," a third girl insisted.

"But I'll tell you a secret," one of the girls whispered in a conspiratorial tone, drawing the others in tight. "In Tokyo, they start much earlier. Some even kiss at thirteen."

For Cho and others like him, the differences between the capital and the countryside are more momentous than the age at which teenagers have their first kiss. To them, Tokyo drains away valuable resources and thus quality of life from the rest of the country. While Americans over ten thousand miles away complain that Japanese companies are overinvesting in the United States, residents of Tochigi, less than one hundred miles from downtown Tokyo, protest that they are not feeling the benefits of Japanese investment at home. The tension between the invigorated cities and the enervated hinterlands, though fresh in the minds of people in Kuzu, Sano, and other rural communities, is one of the least told stories of modern Japan. Tochigi may never achieve the level of development that surrounds greater Tokyo, and may even be better off because of it. Yet for over a century this area and others like it have struggled to move forward while resisting the plight of wholesale exploitation.

This struggle is still celebrated today by honoring the most famous defender of rural lands in all of Japan: Tanaka Shozo, a prominent nineteenth-century parliamentarian from Sano, who led the nation's first grassroots environmental movement against a large mining company for polluting the rivers of Tochigi. After mass demonstrations and national legislation failed to quell the pollution, Tanaka made a special trip to Tokyo in 1901 to appeal directly to the emperor. This effort also failed, and Tanaka was forced to give up his fight.

Today, Tanaka Shozo is to Sano what the *Genjin* is to Kuzu. Just as Kuzu tries to coax tourists with the sex appeal of the "Oldest Man in Japan," so Sano tries to do the same with statues and a new museum to honor the fearless Tanaka and his environmental dream. But as Cho suggested, tourists alone cannot revitalize these areas, and straw heroes cannot keep young people at home. Many in Tochigi have come to realize that they will not flourish again until they are able to reverse the flow of people — and profits — that continue to trickle away, leaving behind a fertile land that has long since blossomed and begun to fade away.

20

BEATLEMANIA AND ALL THAT JAZZ: A TALE OF TWO STUDENTS

If you would form a tree, do so while it is young.
— *A Japanese proverb*

"DO YOU LIKE THE BEATLES?"

I was startled when Takuya, Mr. C's younger son, asked me this question during my first visit to his home. We were eating dinner on a Tuesday night in September, and Takuya was eager to show off his knowledge of Western pop culture.

". . . I love them," he said through a mouthful of rice, not waiting for me to reply. "Sergeant Pepper. Yoko Ono. John Lennon. Pete Best."

"Pete Best?"

"Yes, he was the first drummer for the Beatles, before Ringo Starr. I have an album with his songs on it. Do you want to hear it?"

"Sure."

"But, Takuya," his mother interrupted, "look at the time."

"I know, I know," he said, setting his bowl on the table and gulping down the last of his milk. "I'm afraid I can't play it for you right now. I am late for my *juku* class. Maybe I'll have time later."

He didn't have time that night, and it wasn't until three months later that he finally got around to playing his bootleg album of the former Beatle and showing me Pete Best's picture in the latest edition of *Beatles Monthly Japan*. Takuya was a Fab Four fanatic. He owned two dozen albums, belonged to three fan clubs, and had posters of John, Paul, George, and Ringo on all four walls of his room. But during the many months that I visited his home, I rarely saw Takuya listening to one of his albums or reading one of his Beatles comic books. He simply didn't have time.

Takuya began his day at 7:15 every morning when his Mickey Mouse alarm clock blared from its perch on a speaker above his futon. Fifteen minutes later he would emerge from his room dressed in a clean shirt and shorts from his school uniform, rub his eyes for ten minutes over a breakfast of steamed rice, soy sauce, and raw egg, then hop on his bike for the twenty-minute ride to school, arriving just before the 8:10 bell. For the next six and a half hours, with only a brief break for lunch, he would sit in the second row of his ninth-grade class and listen steadfastly to a series of teachers expounding on everything from calculus to calligraphy, syntax to syzygy, until the clock rolled around to 2:30 and he took his turn cleaning the classroom floor with a rag, stood by his chair for a farewell bow, then dashed downstairs by 3:00 for the start of soccer practice. At 6:30, finally arriving back home, he would rush through the house shedding books and dirty clothes, shovel down some rice with soup and salad, then head out again at 7:00 for an evening of *juku* review. After two more hours of math proofs and pop tests, he would reappear at home after 10:00, dip into the bath when his father and brother were done, then retreat to his room to do school homework and, if he could manage to stay awake, indulge in a spin of his favorite CD.

One Tuesday afternoon in early May I visited Mr. C's house after school and found Takuya lying prone on the living room floor. He was not seriously ill, his mother explained, only exhausted. In an effort to lighten the mood, I decided to perform the nurse's routine I had learned in the hospital. I pushed a thermometer under his arm and asked him various prying questions: how much sleep was he getting (six and a half hours a night), how much homework was he doing (five hours a night), how much television was he watching (an hour a week). When I had finished my examination, I presented him with a small get-well gift of a thousand yen (about eight dollars) and suggested that he take a vacation. He got up off the floor and went instead to study for a social studies exam.

About a week after this episode, I received an envelope in the mail from Takuya. In it he had placed a snapshot of me ministering to his health, with a brief thank-you note attached. "I am feeling much better," he wrote in English. "I am no longer sick. But I am a little unhappy, because I am a Japanese student preparing for an entrance exam."

Takuya's single-minded commitment to school is not unique in Japan, where adolescence is less a time for children to break away from their parents, and more a time for them to accept the burdens that their families — and their schools — place on them. Compared with their counterparts in the West, Japanese teenagers mature slowly. Teachers, for example, call their students *kodomo*, or "children," until they graduate and get their first job, at which point they become *shakai-jin*, "members of society." Parents, meanwhile, allow their children little free time, and push their teenagers to keep their eyes trained on upcoming entrance exams.

The next time I visited the Cherry Blossom home, I decided to find out more about Takuya's private life beyond the public school. I brought a list of questions which I was planning to give to all my students, and at 10:30 one Monday night in mid-May, I sat down with Takuya on the floor of his room and plunged into my survey. Since I already knew about his family life, I skipped the part about his home and jumped to the section on love.

"Which do you prefer, an arranged marriage or a love marriage?" I asked, adopting my best nurse's tone.

"Love marriage," he said.

"Why?"

"Because I want to know the person I marry."

I noted his answer on my clipboard, then proceeded down the list. "Have you ever had a girlfriend?"

He thought for a moment, then answered, "No."

"Have you ever held hands with a girl?"

He thought some more, then repeated, "No."

"Have you ever kissed a girl?"

"Of course not," he said without hesitation. "I live in Tochigi."

Takuya's answers proved typical: only fifteen percent of my students, aged twelve to fifteen, said that they had ever had a boyfriend or girlfriend. Less than five percent said they had ever held hands with a sweetheart, and only two out of a hundred said they had ever kissed, or *been* kissed by, a person of the opposite sex. In the United States one third of teenagers have lost their virginity by the time they turn fifteen.

To be sure, this lack of experience among Japanese is not the result of a lack of interest. Boys and girls love to talk about each other; their music and magazines are full of tales of puppy love and teen *romansu*. It is also not the result of a lack of information. Sex education begins early in Japanese schools, and sixty percent of high school students in one poll said that it is acceptable for teenagers to engage in sex as long as they use contraceptives. For Takuya, however, this day was far away.

"Japanese boys are in love with girls, just like American boys," he said when I asked him why he had never kissed a girl. "But we can only become good friends because everyone else will talk. Frankly, I'm sorry about that."

After sex, we turned to drugs. "Have you ever smoked a cigarette?"

"No."

"Have you ever drunk a beer?"

"No."

"How about sake?"

"Yuk."

"Have you ever sampled drugs?"

"Drugs?" he said.

"You know, marijuana or cocaine."

"Oh no, never. Some friends of mine once told me they smelled paint thinner in a bag, but I don't think that counts."

"Have you ever broken the law?" I asked.

Here Takuya paused to think. He looked over at his stereo system stacked with his collection of Beatles music, then looked back at me. "Yes," he said. "I've broken the law."

"How?" I asked.

"I dubbed a cassette for a friend."

"Is that all?"

"No." He glanced at the floor, twisting his face in apprehension. "I once was really late for school and I ran a red light on my bike."

He wrung his hands in shame at the thought of this major transgression, and I thought to myself: Like father, like son.

With no sex and no drugs, little television and few movies, Takuya's main source of escape was rock-and-roll. But the message of defying authority, so important to the success of rock music in the West, has never caught on in Japan. The musical tastes of my students ranged from the cute to the cloying, from dainty female singers with bows in their hair to clean-shaven boys with skates on their feet. Even popular Western artists, like Madonna and the Beatles, are admired more for their glamorous personalities than for the darker messages their lyrics may contain. I once visited the street in Tokyo where punk teenagers gather on Sunday mornings to "hang out" and "express themselves." These punks sported signature black leather pants, greased hair, and metal-studded shoes, but they didn't seem to flaunt the bitter, antiestablishment slogans of their counterparts in London or New York. Instead, they had formed small troupes, dressed in matching costumes, and were dancing in synchronized routines for assembled out-of-town tourists. Punk had lost its political philosophy and become nothing more than a tame side show to the otherwise bland sounds of Japanese conformity.

Takuya knew this conformity well. After asking him what he would do with a twenty-fifth hour in the day (sleep) and what he would buy with one hundred million yen or $650,000 (a new stereo), I posed the final question in my survey.

"What is your goal in life?"

I expected him to pause and consider his answer. I thought he would think through all the messages he had learned in his life — the hopes of his parents, the advice of his teachers, the signs that hung on the walls of school like the Ten Commandments of youth. In fifteen years, he had learned to be hard on himself and soft on others; to defer to his elders and assist his juniors; to go on green and stop on red.

But Takuya didn't pause at all. As soon as he heard my question, he looked up at me and said, "My goal is to go to high school."

The voice of prudence: a rising son.

I knew a girl in Sano who didn't like the Beatles. She hated Madonna as well. Her name was Aya, and she preferred jazz. Aya's dream was to pursue her love of music and sing solo someday in a Tokyo nightclub. Aya also did well in school — especially in math and English — and her teachers were urging her to choose a more conventional career path. "I really love music," she told me with a sigh, "but my teachers say I should do something more secure, like teaching or nursing."

Almost every day for several weeks in the spring, Aya came by my desk at school with new questions that she had thought about the previous night. "What is the life worth living?" she asked. "Is death final for us?" She seemed eager to make a decision.

In some ways Aya was a traditional, "girllike" girl. Her father was a tailor, her mother a housewife. She too wanted to get married, have children, and start a home. Aya prided herself on her formality. "You don't know how to pour a cup of tea," she told me bluntly one afternoon. "I will teach you." And she did, instructing me how to grip the teapot with my thumb and middle fingers, freeing my index finger to hold the top in place.

"I bet you like to cook," I said when we had finished.

"I cook very delicious," she said. "My mother is a good cook, and she taught me."

Yet Aya knew she would not be happy pouring tea all day in her father's shop or tending patients in the Welfare Hospital. "I want to have children," she said, "but is that enough? I want to know what makes people happy."

Over the next several weeks Aya came to the preliminary conclusion that there were two kinds of happiness in the world: "universal happiness," in which people fit in with a larger plan, and "detail happiness," in which they follow a personal plan. The question she faced was how to balance these conflicting goals.

"I want to have a husband," she said, "but I am not very beautiful. Maybe I will not make a good wife.

"I want to sing on television," she countered, "but maybe I am being too selfish.

"My parents tell me to ask questions, but my teachers tell me I should just study for the tests."

Instead of encouraging the Ayas of Japan to go in search of their dreams, Japanese teachers tend to put students — especially girls — in their place, and teach them to keep their expectations in check. As a result, most children know by the end of junior high school what direction their lives will take. My students were surprisingly unromantic about their future plans. When I asked them what they wanted to be in the future, several students wrote down truly "dream jobs" like professional soccer player or film director, but most made prosaic choices like bank teller, bus driver, or beautician. For all the talk in schools about merit and equal opportunity, most Japanese students have an air of fait accompli about them. Even though Japan has increased its standard of living dramatically over the last forty years, less than half of my students said they expected to earn more money than their fathers. Clearly these children were not being taught to reach for great challenges; instead, they are taught to get ahead by going along. Like miniature bonsai trees, Japanese children are pruned while they are young so that when they ripen later in life, they do so in harmony with the world around them.

About a week after Aya sat at my desk contemplating the question "What makes people happy?" I arrived at school one morning to find a postcard from her in my chair. On the front of the card was a nativity scene, with Mary, Joseph, and the baby Jesus; on the back was a note in English: "I write you a postcard. My postcard has strange message. Are you vigor? I am vigor, because I decided that happiness can be found on a detail level. In the future, I try sing."

The voice of hope: a rising daughter.

While most students, like Takuya, will emerge from school with their futures clearly marked and their branches neatly trimmed, a growing number, like Aya, will want to take a chance away from the group and try to blossom by themselves. Japan's next great challenge will be to broaden its ideal of the model citizen to include this new kind of student — one who wants to play as well as work, to have pride in herself as well as her country, to grow into a full-sized tree. To scorn these students because they resist the orthodox mold would be the greatest misfortune of all. For as all of us in Sano would soon be reminded, even those who live outside the circle need to feel at times that they still belong.

21

OUTSIDE THE CIRCLE: THE INVISIBLE CLASS

At one time
Alone, in a land of strangers,
I denied my childhood home.

Peter thrice denied his Lord.
How many more times did I deny
My humble place of birth?

— *Maruoka Tadao, "Why Flee?" 1958*

KENZO SAIKAWA SAT UP *quickly in bed — yet another fitful, sleepless night. Peering out at his crowded room, he groped at the floor in search of his clock. It was 6:15, an hour before the alarm would sound and his mother would peer around the door. Outside, the first blush of the mid-May sun had already begun to stir the sky. Kenzo had little time to waste. He reached for his creased black pants but paused for a moment, lifting his diary from where it lay on the sheets and stashing it beneath his bed. Someday, it alone would speak the truth.*

Kenzo dressed in his formal junior high school uniform. He carefully buttoned the jacket to his chin and removed his bookbag from the back of his chair. The night before, his mother had sat in that same chair as he told her about the boys at school. They had been teasing him for almost a year, he said. At first it was only little taunts. "Kusai," they said, "you stink." "Baikin da, you're a germ." But later some began to hit him and push him down the stairs. The week before, one of the boys in his basketball club had asked him to steal a box of candy from the 7-Eleven.

"What was his name?" his mother asked.

"Takajima," the boy said.

"Takajima?" she gasped. "But I thought he was your friend."

"It's not true!" the boy screamed. "He's not my friend. That's just what my teacher said. Nobody listens to me. Everybody thinks he's my friend, but he's not. He hates me. He says that I am different, I look funny, and I don't deserve to be in school."

His mother assured him that this wasn't true. He could attend any school, have any job, or say anything he wanted, she promised. His teachers had told him the same thing. But his classmates said the opposite. They teased him about his name — Saikawa — and the exclusive classes he attended after school with several other students. "You get all this special treatment," Takajima had chided him many times, "but with me you are not so special." Kenzo lay crying on the bed when his mother finally left the room: some hurts could not be silenced.

The boy hurried to leave the house before seven o'clock, when he knew his parents would wake up. He tiptoed down the hall, slid on his graying sneakers, and slipped quietly out the front door. Once on the street, he turned in the direction of school.

Several members of the girls' track team were jogging around the playing field when Kenzo passed through the gate at 7:15. No one seemed to notice the small ninth-grade boy making his way across the track toward the gymnasium. Kenzo had never attracted much attention at school. He was mostly silent. The teachers all thought he was "quiet."

Kenzo did not stop at the gymnasium, where his fellow members of the basketball club were holding a morning practice. Instead, he went straight into the school building, left his sneakers at the door, and walked in stocking feet up the stairs to his classroom on the third floor. The room was empty when he arrived. The lights were off. He dropped his bag on the floor beside his desk and took his seat in silence. Although he couldn't see the students outside, he could hear their voices through the open window and again as they echoed down the vacant hall.

His mind was cluttered, and the voices seemed to be creeping *closer. He was alone, but somebody seemed to be calling out to him. He heard the voices:* kusai, baikin, hinin — *dirty, stinking, doglike boy. He pictured them teasing him during lunch and on the way home from school. Takajima was the worst. One day he would call out dirty names; the next day he would act really nice. Kenzo thought everybody knew about this. They could look into his eyes and know he was alone. They could look beneath his clothes and see the scars where some of the boys in the basketball club pressed burning cigarettes against his arms. They could tell all this, and yet they let it continue.*

The voices were growing louder now. Perhaps somebody was coming. Kenzo reached into his bag and removed a folded sheet of paper. He placed it on his desk and hurried toward the back of the room, where the sliding glass doors opened onto the balcony. From here he had a view of the entire school and the joggers down below. Kenzo knew they couldn't hear him, so he didn't try to speak. He knew they wouldn't listen, so he didn't try to scream. At this moment — as always — he spoke only to himself.

At 7:25 in the morning of a bright Tuesday in May, Kenzo Saikawa walked alone to the edge of a third-floor balcony at Sano Junior High, stepped over the concrete rail, and reached for a sound that would silence the words that echoed in his heart: "You are different. You don't belong . . ."

The sudden death of Kenzo Saikawa stunned the entire community of Sano Junior High. For several days afterward, through a series of memorial services, discussions, and counseling sessions, the school grieved the loss of one of its own and tried to understand what had precipitated this act of self-destruction. From a series of these discussions we were able to piece together an account of Kenzo's last hours.

About a week after the suicide, I sat with Denver at a local teachers' hangout, eating sushi and drinking watery beer, as he explained to me that for much of Japanese history, while samurai warriors freely roamed the country battling one another for territory, a special class of menial workers known as *burakumin,* or hamlet people, was confined to ghettoes and legally isolated from the rest of society. Though modern laws have eliminated this segregated system, its legacy still pollutes the minds of some children today who have inherited the prejudices of the past.

"I don't know if it is good to tell foreign people," Denver said, "but this is a problem in Japan. The distance between these people and the rest of the population is the same as that between Japanese and foreigners. They are treated like aliens."

Denver seemed fatigued, his tie hung loose around his neck, and his hair was disheveled from running his fingers too many times across his scalp. He was weary from talking with students, frustrated at their parents, and tired of sustaining the dam of silence that had risen around the school. He took a napkin from the bamboo box on the table and quickly sketched a pyramid, which he split into four separate tiers.

"In the Tokugawa era," he said, "Japanese society was divided into four categories: the samurai were at the top, followed in order by farmers, artisans, and merchants. But below those people was another group." He drew an oval beneath the pyramid and shaded it black like a puddle of mud. "They were called the *eta*, the 'people of filth.' Everyone studies the top four groups, but not everyone knows about the *eta*. It was not until I became a teacher that I became fully aware of this exclusion."

The people confined to this group were mostly those who worked with animals — butchers, tanners, and leather workers. Buddhist taboos against killing animals, combined with Shinto fears of uncleanliness, conspired to isolate this entire class of laborers.

"They looked the same as other Japanese," Denver explained. "They even talked the same language. The only difference was their work. The samurai lords assigned them special names and forbade them from walking, talking, or marrying with people from other classes. The *eta* weren't exactly slaves, but they certainly were not free."

Denver looked around to make sure no other people were

listening. We were alone in the bar except for three men drinking on a raised platform in the center and a couple eating dinner at a freestanding table in the corner. A light breeze blew through the navy blue curtains that hung over the open door, jingling a chime that dangled over the bar. The master rolled a piece of sushi at the counter and tallied the cost on a tablet by his side. Denver refilled my glass with beer and continued to talk.

The legal segregation continued until the 1890s, he said, when Japan's first constitution officially outlawed all class distinctions. But the discrimination persisted. The *burakumin*, as the *eta* came to be called in the twentieth century, could still practice only certain professions, and unlike the displaced samurai, they received no money from the state to help them integrate into the mainstream of society. Although they looked, talked, and dressed like all other Japanese, the *burakumin* still lived apart. They remained, in effect, an invisible race.

"Many of these people changed their names and moved into new neighborhoods," Denver said, "but, of course, everybody still knew. Each person in Japan used to have a personal record at the courthouse which listed his relatives, birthplace, and occupation. Nobody dared get married without first checking this record. Japanese people can be very wary, you know."

At the close of the Second World War, a new emancipation effort emerged. A group of *buraku* leaders, calling itself the Liberation League, began to focus attention on what it saw as the best avenue for salvation: the schools. The group pressed the government to meet two key demands: first, teach all students about the problems of discrimination; and second, hold special classes for minority students after school to help them prepare for entrance examinations and thus gain admittance to more competitive schools — a sort of school-sponsored *juku* for *burakumin*.

"The situation is getting better," Denver told me. "Unlike twenty years ago, almost all *buraku* students attend high school, and some continue to universities. Also, we have eliminated the system of personal records at the courthouse. Our ancestors created the problem; it is our duty to solve it. The afterschool classes are one way."

Yet for those four or five students in Sano who went off twice every week to these sessions, the benefit gained by specialized attention was often overshadowed by the stigma attached to attending. The other students quickly realized that those who attended these meetings were somehow different from the rest of the school. The *buraku* students were often hazed in return as dirty, filthy, children of dogs. And as any person who has lived through junior high school can attest, such curses are as sharp as swords.

"It's a shame about that young boy," the sushi master said as he came to clean our trays at the end of the meal. His lips were puckered around a cigarette, his eyes squinted from the smoke, and a hint of white beard peeked out from his chin. "I sure hate to see a student —" He stopped himself suddenly in midsentence and tilted his head toward me. "Does the *gaijin* know what happened?" he asked.

I told him that I did.

"Where does he come from?" he said to Denver.

"The United States," I answered.

"America," he said. "That's great. Me, I don't speak English. It's a shame he doesn't understand Japanese."

"But I do understand Japanese," I said.

"Oh," he answered, now turning toward me. "I guess you're right." He stacked the empty trays atop his right arm and took a drag of his cigarette with his left.

"I'll tell you something about Japan," he said, wagging his cigarette in my direction like a teacher gesturing with chalk, "compared with America, we seem very equal here. But that's not completely true. There are a lot of things you don't see, and a lot of problems you don't hear about. Me and *sensei* here have the same heart, we can understand each other, but there are a lot of people in this country who don't treat others right. This is the shame of Japan."

After the master had returned to the bar, Denver laid another napkin on the lacquer tabletop and began to draw what looked like a solar system, with a large sun in the center and a group of smaller planets around the outside.

"A long time ago," he said, "this was the shape of Japan. We call it *bushidō*, the 'Way of the Samurai.' In the center was the lord — the chief of a tribe — and around him was a band of retainers."

"A *kumi*," I said.

"That's right," he said. "Above all, each person in the group was required to remain loyal to the clan. To make sure this happened, the samurai developed a special code: if any member of the group committed a shameful act, then the lord — the leader — must take the blame. He must seek an honorable end."

In the Way of the Samurai, an "honorable end" meant a ritual form of suicide known as *seppuku* or *harakiri* — literally, "cutting the stomach" with a sword. Japanese literature is full of melodramatic legends such as "The Tale of the Forty-seven Samurai," based on the true story of a band of retainers who avenged the death of their lord by assassinating his rival. The royal court, forced to respond to the murder, ordered the retainers to commit public suicide, thus fulfilling the letter of the law but still allowing the warriors to rejoin their leader in the afterworld. When I saw this story performed in the stylized kabuki mode in Tokyo in early spring, the audience cheered its uproarious approval as the forty-seven men dressed in brilliant purple kimonos lined up across the snow-covered stage and disemboweled themselves en masse.

"Of course most modern Japanese can't imagine doing this to themselves," Denver said, "but suicide is still an honorable way out of a desperate situation. Just as a samurai lord killed himself in the past if his retainers committed a shameful act, so a section chief or leading politician might kill himself today if some member of his staff is caught breaking the law."

The effects of this legacy are dramatic. Japan has an annual suicide rate that is fifty percent higher than in the United States and places Japan among the leading countries of the world in the percentage of self-inflicted deaths. But while adult suicide is common in Japan, teenage suicide is not. In 1987, about 1,700 Japanese between the ages of fifteen and twenty-four killed themselves, at a rate that was fifty percent less than in the United States. In a broader comparison, Sweden, Switzerland, Hungary, and El Salvador all have higher teenage suicide rates than Japan.

A prevailing myth about youth suicide in Japan is that those who take their own lives do so because of "examination hell." The anthropologist Thomas Rohlen conducted a survey, however, which revealed that youth suicides were distributed throughout the entire year, not just at test-taking time, and that the reasons for the deaths ranged from a lack of motivation to study, to a problem with the opposite sex. As the case of Kenzo Saikawa shows, the death of a student is often the result of tensions among children themselves, not just between children and adults.

Although Japan has a long and honorable tradition of self-inflicted violence, it does not have a legacy of violence against others. The number of people killed by handguns in Japan in a recent year, for example, was 35, compared with 9,104 in the United States. Japanese people are justly proud of this tradition. Their country may be small, they brag, but it is "safe."

Like most other tenets in the canon of Japan, this one is changing. Japan's recent economic success is helping to create a new generation of youth that does not blindly follow the old rules. In 1987, for example, teenagers committed almost half of all crimes in Japan, including shoplifting, burglary, and car theft. Soon after the suicide at Sano Junior High School, I clipped a page from an English-language newspaper in Japan. Side by side on the top of the page were two banner headlines: "POLICE ARREST 16-YEAR-OLD IN STABBING DEATH OF TEACHER" and "JUNIOR HIGH TEACHER NABBED IN FATAL STABBINGS OF RESTAURANT OWNER, WIFE." These two incidents were not directly related, yet in a larger sense they were. Violence seems to be gaining currency in the lives of Japanese children, and many people fear that schools may be encouraging this trend.

Although disciplining students with a whip, a stick, or a slap across the head was outlawed after the war, almost three quarters of all Japanese teachers admitted in a recent poll that they still used corporal punishment. In Sano I regularly saw teachers force students to sit on their knees for long periods of time; other teachers slapped offenders on the head. Teachers who occasionally pushed students, kicked them, or shoved them away were common. But often the line is blurred between gentle reminders of who is in charge and more serious efforts to inflict pain on students. In a notorious 1986 incident, a teacher in the south of Japan beat a student to death for bringing an outlawed blow dryer on a school excursion. At least one teacher I knew in Tochigi kept two hollow bamboo sticks under his desk for more serious transgressors.

While it would be an exaggeration to say that Japanese students go to school in an atmosphere of violence, it is fair to say that schools generate a high level of stress in the form of pressure to conform and comply with the rules. This invisible violence in schools, like "white noise" in cities, lingers in the air, constantly reminding students of the threat of force that surrounds them at all times. A growing number of children, called "school refusers," have responded by staying home. Other students take out their anxiety on one another in the form of teasing, taunting, or bullying.

"What actually caused the boy to kill himself?" the master asked Denver from behind his cash register when we came to pay the check. He tabulated the bill on a wooden abacus, then punched it into the register.

"He was a member of the basketball club," Denver said. "Some of the boys had been teasing him, beating him, even asking him to steal. He wanted it to end."

"*Ijime*," the master sighed. "Just as I suspected."

The *ijime* pattern of student-on-student violence has become fairly well established. It begins with minor taunts — "You stink," "You're a germ," "You don't belong in this school." Then it moves on to petty crime — forcing a student to steal candy from the 7-Eleven or a pack of cigarettes from the railroad station. And it often escalates to the level of physical abuse — cigarette burns on the arms or punches to the head. Much of this goes unseen by the teachers, who know it exists but do not actively try to stop it. Most of the victims never speak out. Instead, they learn to live with their torture as just another price for being different from their peers. Occasionally, however, a victim will lash out, and the consequences are dramatic: a seventh-grade student beats himself to death with a hammer; an eighth-grade girl hangs herself in her home; a ninth-grade boy jumps off the balcony of his third-floor homeroom class. These children could think of no way to escape, so they finally decided to join the others and persecute themselves.

"There is something else you should know," Denver said as we walked out of the restaurant into the moonless night. The warm, moist air seemed to drip from the trees like molasses from a spoon. The breeze had died down again.

"What is it?" I asked.

"You should know that the samurai spirit still exists in Japan."

"You mean in companies, and offices . . ."

"I mean in schools," he said, pausing to lean against the hood of his car. "In each class, students are like a group of retainers who join beneath the banner of a lord. When one of those students dies like this, someone must take the blame."

"A student?" I asked.

"A lord," he said. "The leader must always bear the shame and seek an honorable end."

22

P'S AND Q'S AND ENVELOPE BLUES: A JAPANESE WEDDING SPECTACULAR

A wedding, be it large and elaborate or small and simple, is one of life's most important occasions — beautiful, meaningful, and traditionally a couple's day of days.

— *Emily Post,* Etiquette

ONCE, DURING MY EARLY MONTHS in Sano, I wrote a letter to some Japanese friends while I was at my office. When I had finished, I put the letter into an envelope, copied the address of my friends on the front, and gave it to Arai-*san*, the affable "office lady" who daily gathered the mail.

Several days later, Mr. C approached my desk with my letter in hand. "Mr. Bruce," he said in a low voice, squatting beside my chair as he did when he had important matters to discuss, "I'm afraid we cannot mail this envelope."

"Why not?" I asked in a similar hushed tone. I assured him that the letter contained important office business.

"It's not the letter," he said, "it's the envelope. You have not prepared it correctly."

"But the address is accurate on the front," I protested, "and I wrote the return address of our office on the back. Can you not read my writing?"

"Oh no, your writing is very beautiful — more lovely than mine," he said in one of those fatuous compliments that usually warned of something harsh to come. "But you have forgotten a very important detail. You left out the character for *sama* [a more formal, written version of the honorific *san*]. I'm sorry, but we cannot mail this letter without it."

Was this a joke, I wondered — a parody, perhaps, of the Japanese obsession with detail? Would the authorities in this office really not mail my letter without the Japanese equivalent of "Mr." or "Mrs." scripted on the front? Did they really check every letter that passed through the mail bag to make sure that all the names were anointed and all the *kanji* were crossed? The truth, I realized, was that this was no joke. The senior secretary of the Ansoku Education Office of the Tochigi Prefectural Board of Education had delivered my mislabeled letter to her section chief, who had conveyed it in turn to *my* section chief, who had passed it finally to my boss, who had dutifully come to inform me that the Japanese government refused to spend sixty yen to mail any envelope that did not contain the proper appellation of respect.

I thanked Mr. C for his advice, apologized for the inconvenience, and told him I would solve the problem. But instead of just adding the missing character, as any humble civil servant would have done, I vowed to prove that my officemates had grossly overreacted. I took the errant letter to the post office, purchased a stamp, and mailed the envelope myself — *sans sama*. Any friend of mine, I thought, would not be offended by this trivial lapse of etiquette.

The next time I visited these friends — a middle-aged couple in Osaka whom I had known for some time — I related this story to them, expecting us all to share a hearty laugh. But when I reached the end of my story, my friends didn't laugh. They didn't even titter.

"Mr. Bruce," they said with utter sincerity, "your boss was right. We always know when we get a letter from you, because you never address an envelope in the proper way. Form is very important, you know."

Envelopes, as I learned the hard way, are more than mere

packaging in Japan. They are more than simple wrappers that protect a private letter and are later thrown away. As a school uniform defines a student or knickers a mountain hiker, an envelope actually becomes a part of the message itself. "In Japan, the package is a thought," wrote the philosopher Roland Barthes. Within minutes of reprimanding my poor form, my friends led me to a special drawer in their home which they reserved exclusively for new envelopes. Inside they kept containers for every occasion — from births to deaths, from New Year's gifts to mortgage payments. Some were wrapped in ribbons of red, while others were garnished with silk cherry blossoms. They even had a special envelope for the tooth fairy.

The last, and most elaborate, package they drew from their drawer actually consisted of two wrappers in one. On the inside was a white sheet of paper, folded twice to conceal its secret contents, and on top of this slid a thicker slip of paper which was sheathed in red and white twine, knotted around the midsection, and adorned with sprigs of pine.

"This is the most precious kind of envelope in Japan," my friends said as they handed me this paper bouquet that seemed more suited for framing and hanging than licking and stamping. "We put a crisp yen note in the inner fold, tuck this into the outer sheet, wrap both sheets inside a silk handkerchief, and give it to a bride and groom on their wedding day. This is our Japanese custom."

In early June I got the chance to put my new envelope expertise to a test and spend a Sunday afternoon away from the tensions of Sano: I was invited to attend the wedding of my new friend Hara, the banking tycoon from Tokyo, and his bride, Emiko, from the "Up River" Real Estate Company. Cho and I drove together to Tokyo and arrived several minutes before noon at the canopy-covered entrance to the Mikado Hotel, across the tree-shrouded moat from the estate of the Japanese emperor. Upon arrival we were obliged to give a present — in cash. This courtesy contribution, currently running at 20,000 yen (or $150), defers some of the cost of the ceremony, but mostly pays for the gifts that the couple is obliged to give the guests. For my gift I got to witness the marriage ceremony, have my picture taken, savor a five-course meal, and — according to the complex calculus of gift-giving in Japan — receive *four* presents in return: a chocolate cake, a bag of rice, a pair of crystal goblets, and an economy-sized bag of dried fish shavings. Later I also received a box of macadamia nut chocolates from the couple's honeymoon in Hawaii.

After registering at a table in the lobby and handing over our envelopes, we were hurried up a winding staircase and ushered into a cramped waiting room where the two families sat facing each other in resolute silence, like rival diplomatic delegations across a negotiating table. The groom's family sat to the right in three even rows of ten people each, and the bride's entourage sat directly across in a similar, formal phalanx. Like most modern couples, Hara and Emiko held two ceremonies: a private religious service for family members, followed by a large reception for friends, colleagues, and obligatory guests. Cho and I, the only friends invited to the private ceremony, were placed between the two families on a third side of the square, directly across from the bride and groom.

Hara, his face somber, wore a black silk kimono with embossed family crests, wrapped by a pure white sash, while his bride wore an ornate white kimono with winglike sleeves embroidered with peacocks and birds of paradise. Her usually beaming face had been powdered over with white cake make-up and quieted with a submissive smile drawn with red lipstick. On her head Emiko wore a crowning, boxlike white hat that rose ten inches from the top of her hair and hung out over her eyes like a Greek pediment. This elaborate headgear, known as a *tsunokakushi*, is designed to hide the "horns of jealousy" that all new wives are expected to sprout.

"Would both the mothers please stand," the hotel attendant announced from the front of the room, "give your name, and your place of birth. Would the grandparents please rise . . ." I felt as if I was on jury duty, watching witnesses being sworn in. Ten minutes later, with the formal introductions completed, we moved two doors down the hall to a more traditionally appointed Japanese room for the nuptial ceremony itself. Black wooden chairs lined three sides of the room, spaced evenly across the lustrous wood floors that shone like melted butter. Narrow screens with charcoal strokes of bamboo leaves leaned in from the ivory walls. The guests took their places along the outside of the chamber; the bride and groom stood on virgin tatami mats in the middle of the room before a miniature Shinto shrine. The freestanding mahogany shrine had all the poise and strength of an antique armoire, with a carved frontal piece, several shelves holding gems and fruit, and a mirror that reached up from behind the altar and was trimmed with garlands of green summer leaves.

A white-robed priest emerged from a door behind the shrine, made some opening remarks, and then removed a white sheet from atop a large straw keg of sake which was resting before the altar. Moving slowly, he blessed the couple with his hands, waved streams of white paper over their heads, and poured some clear sake into a cup at their feet. The bride knelt down, lifted the scarlet cup to her lips, then offered it to the groom. When he had finished, the priest refilled the cup. The groom drank first this time and tendered the cup to his bride. Finally, each of us in the wedding party retrieved a cup from under our chair and waited for the priest to fill it with rice wine. The priest returned to the altar and clapped twice to summon the gods; we all bowed as one, then drank from our cups. What wine has joined together, let no one put asunder.

Throughout the short ceremony, the priest gave no indication of having met the couple before. He read standard prayers from a photocopied sheet of paper, and at each stage of the ceremony he leaned over and whispered instructions into the groom's ear: "Put on the ring . . . Bow! Start walking out . . . Now." The Shinto element to Japanese weddings is relatively new, added only after Christian missionaries poured into Japan a century ago and introduced their religious ideals of marriage. Before that time, marriage was viewed as a secular union between two families. Nevertheless, the ceremony itself seemed devoid of any religious ambience. There was no music except for a few tape-recorded reedlike sounds piped in over scratchy speakers, and no prayers were uttered by the bride, the groom, or either of the families. The only other people who participated in the ceremony were an older couple who sat behind the wedding pair for no apparent reason.

"What were they doing in the middle of the room?" I asked Mr. River Up after the ceremony had ended and we were having a group picture taken.

"They played the part of the *nakōdo*," he said, "the formal go-between."

"But I thought Hara and Emiko met through friends," I said. "They didn't have an arranged marriage. They were a love match."

"That's right. But the hotel told us we had to have a *nakōdo*, so we asked the neighbors to sit in."

After the modest ceremony, we moved back downstairs for the reception and joined 150 of the couple's assorted friends and co-workers in the grand ballroom of the hotel. Far from the refined elegance of the Shinto chapel, this room was about as understated as the palace at Versailles. Mauve velvet curtains swooped like condors from above and draped the sides of the gilded shutters that covered nothing but wall. Buxom chandeliers bobbed from the ceiling like frilly hoop skirts at an antebellum Southern ball. The round tables were heaped with twinkling topiaries; satin ribbons tied to the top dribbled over the edge of the linen tablecloths and dragged along the floor. Cho and I found our assigned seats, along with Hara's other college mates, and sat in front of silver platters brimming with cold meats, sliced fish, and pickled vegetables — all covered with plastic, so we could look but not touch.

The reception itself was so elaborate that it required a five-page printed program and a rented emcee to narrate the show. In the beginning was the word — speeches, to be precise. The happy couple, now changed from their ceremonial garb into a less regal but still formal pair of kimonos, marched down the center of the room and took their places on a raised platform for everyone to see. The acting matchmaker then assumed the microphone and proceeded to recapitulate the life story of the groom. "Hara-*kun* was born in a small wooden house on . . . As a boy, he obeyed his parents and worked long and hard at . . . He attended elementary school at . . . In junior high school he was a member of the . . . His favorite class in high school was . . ." The only thing omitted from this exhaustive résumé was a list of his former girlfriends. The matchmaker's wife then did the same for Emiko, followed by more character testimonials from the groom's boss, the bride's sister, several high school friends, and even Emiko's junior high school homeroom teacher. Cho, in his capacity as honored university elder, gave a short address about the virtues of traveling with long-time friends and the difficulty of traveling with Hara, who always came away with the girls. The audience applauded politely, like white-gloved guests at high tea.

After the last formal toast, "To the wedding of Toshiaki Hara and Emiko Kawakami," we were finally able to open the wine and remove the plastic wrapping from the food, which had been growing steadily more appealing on the table before us. With the start of the meal, tuxedoed waiters hurried to our sides, dishing out portions of French onion soup, Caesar salad, and lobster not-quite-Newburg made with processed cheese, since the Japanese don't care much for *fromage verité*. The real show, however, was on the stage. After a pause long enough to admire the meal but not long enough to eat it, the lights were dimmed and a movie screen descended magically. Through sips of soup and sweet German white wine we watched a slide montage of childhood pictures of both the bride and the groom, set to the rueful tune of "Time in a Bottle."

When this high-tech montage was finished, the emcee called for the lights to be blackened completely, and his band of loyal retainers rolled out a small computerized machine about the size and shape of a grocery cart. As the speakers whined with Lionel Richie singing "Endless Love," tiny beams of red, white, and green light burst from the box with a shower of brilliance and began to dance in the dark. "It's the Mikado Hotel's Laser Light Extravaganza," the emcee wailed as the light beams outlined frolicking butterflies and dancing hearts on the wall in the back of the room. "TOSHIAKI AND EMIKO — TOGETHER HAPPY ECSTASY," the laser scripted in classic Japanese-English. "LOVE IS 4-EVER." The crowd shrieked its approval. "Isn't it spectacular!" said a woman at a nearby table. "It's better than the Magical Light Parade at Tokyo Disneyland." I thought for a moment I had found the ultimate trophy of Japanese technological ascendancy: a portable electronic box that painted multicolored sea gulls on hotel walls to the squealing delight of hundreds of guests.

But we were not through yet. Just as the laser show drew to a close, a spotlight reached out from the darkness like a shining sword and revealed the bride and groom, now dressed in chiffon wedding dress and tuxedo, being lowered into the room inside an eight-foot-tall white-picket gazebo suspended from the ceiling. As this Cinderella-like coach touched the ground, the crowd oohed and aahed and two dozen children wearing pinafores and sailor suits stepped forward to greet the newlyweds with pink carnations in hand. Emiko kissed the children on the cheek; Hara shook their hands; and my *go-con* companion Prince Charming, in his capacity as cameraman-at-large, rushed forward to capture the scene on his portable video camera. Moving back toward the front of the room, the happy couple mounted a small round stage alongside a three-tiered pink wedding cake, which was festooned with lacy icing, supported by plastic Ionic columns, and topped with a Caucasian wedding couple underneath a canopy. The lights dimmed. The crowd hushed. The master of ceremonies announced into his microphone, "This is the climax." The bride and groom brought down the knife together with all the ardor of an aspiring samurai, and suddenly the stage began to rotate, the cake began to shake, and pink smoke came billowing out from beneath the lowest tier. As the tape-recorded violins soared to the crescendo of "Love Is a Many-Splendored Thing," white spotlights drowned the stage and the entire platform began to rise on the shoulders of three hydraulic beams, like a UFO taking flight. I held my breath, thinking for a moment that the cake was going to lift into the air on a web of red and white laser beams. Yet the crowd could contain itself no longer. Roaring their approval, the guests jumped to their feet in riotous applause and swarmed the swiveling cake with an arsenal of flashbulbs and dessert forks.

So much for the myth that Japan is a land of understated elegance.

When I went with Hara on our failed *go-con* in January, he told me about a new era of love in Japan — or at least in Tokyo — in which sex comes first and then comes marriage. In this new romantic age, he said, boy meets girl, propositions girl, then takes her home to bed, "just like in America." The apogee of this modern love is the modern wedding. More than any other event I witnessed, this reception showed how material wealth is changing the lives of the "way out, but classic" generation. In the past, weddings were more sober affairs, befitting the *kejime* of moving from one stage of life to the next. Mr. C, for example, was married in the shrine I had visited on New Year's Eve. But weddings today have become showcases for wealth and gadgetry. The average Japanese couple spends $53,000 to get married — about half from the cash contributions and the rest from the parents of the bride and groom. The high cost of tying the knot includes not only the price of the wedding and reception but also $1,000 as a finder's fee to the real or stand-in *nakōdo*, $2,000 for photographs, and an estimated $10,000 in dowry money exchanged between the two families. Since a wedding is a milestone for an entire community, many families are willing to splurge to enhance their position among their friends and neighbors.

As I witnessed the parade of these high-priced toys, from laser-light moonbeams to hydraulic-powered cakes, it occurred to me that some money-conscious Japanese may have fallen victim to their own brand of "conspicuous consumption" — spending enormous sums of money for extravagant tokens of wealth and status. If anything, money has allowed the Japanese to explore their wildest fantasies, especially of romantic love. Magazines like *Seventeen* and *Jump!* — modeled after prototypes in the United States — have saturated Japan with their dreamy tales of Western love. Comic books, soap operas, and bubble-gum pop music all urge young people to spurn the formulaic methods of courtship which their parents pursued and set out to find the perfect match for themselves. Young people have taken this foreign custom and carried it to extremes. The Japanese imported Valentine's Day from the West, for example, and promptly added a parallel holiday one month later called White Day. On February 14 girls give boys gifts of chocolate, and on March 14 boys return the favor with batches of homemade cookies. Hikaru Genji — a popular musical band of beatific roller-skating teenage boys named after the gallivanting hero in *The Tale of Genji* — received an astonishing eighty tons of chocolate one year from weak-kneed junior high school girls all across Japan. Even though marriage is still primarily an institution in which the wife is expected to show duty toward her husband, a glitzy wedding can provide a temporary escape from this reality. On its wedding day, at least, a couple is elevated from the bonds of obligation to the realm of *romansu*.

Yet in the course of the entire reception, no speaker made any reference to the future happiness of the bride and groom. No person even mentioned their "love." Besides exchanging rings — another Western adoption — the couple did not touch and, as far as I could tell, did not even look at each other throughout the entire four-hour ceremony. Not until the end of the day, after we had all filed out of the hotel with our shopping bags brimming with chocolate cakes and crystal glasses, did the newlyweds emerge, in their fourth change of clothes, to reward us all with their official "first kiss." As I watched, I remembered Hara telling me on the evening of our *go-con* that he would ask his wife to quit her work as soon as they were married. He wanted modern love, all right, but with a traditional wife. Despite all the bells and whistles, the new age of love imported from the West has been unable to bridge the age-old gap that separates men from women, even those who are married to each other. The wedding itself was simply an "event," with little spontaneity, little emotion, and, despite the laser valentines and sentimental soundtrack, little heart. After all the excitement was over, the wedding reminded me of its own envelope — an elaborately crafted package stuffed with brand-new money.

23

A PASSAGE TO DISNEY: THE ANNUAL SCHOOL EXCURSION

Mickey is the natural leader and the smartest of all the Disney crew. He is cheerful, sensitive, warmhearted, and generally likes people.
— *A quote on my office trash can*

JUNE BROUGHT FRESH WINDS INTO SANO. In my neighborhood, baseball season began in earnest. The boys across the street decided after great consultation to place home plate just in front of my mailbox, which meant that home runs landed in the cemetery behind my building and foul balls in my kitchen. Around the corner, the Hotel Sunroute was so swamped with guests that it began valet parking in the evenings and put a concierge with a walkie-talkie in the lobby to direct guests to the French bistro on the ground floor, the outdoor beer garden on the roof, or the main ballroom upstairs to the right, where the surge of lavish weddings on weekends had increased to three or four a day. Just down the street, the Jusco Department Store sponsored a week-long sidewalk sale featuring cotton sheets and "towel-kets," lightweight terry blankets popular during humid summer months. And across town at the Cherry Blossoms' home, Mr. C spent several weeks taking down the thick metal shutters from the windows and placing corrugated plastic over all the doors, while Mrs. C stored the wool kimonos in cedar chests and began burning incense in every room to reduce the dampness inside the house. The reason for this flurry of activity was the coming of the rains, which arrive every year at the end of June to sadden the children, hearten the farmers, and rekindle the grain of the gods.

In school, students marked the change of seasons by doffing winter jackets and donning short-sleeved shirts, but their minds were focused on a coming attraction of a different sort. Classes plotted for weeks; teachers fretted for months; the television news anchors even announced on the air when the season officially arrived. Every year just before the midsummer rains, junior high schools from coast to coast load their students onto trains and buses, read them their rights as apprentice citizens, and send them out into the world for the annual rite of passage known as the school excursion.

At Sano Junior High, each class took a trip. The seventh graders visited Tokyo Disneyland; the eighth graders made an overnight hiking trip; and the ninth graders took a three-day excursion to the ancient capital of Kyoto. Even though these trips were great fun for the students, the teachers emphasized that they were also educational experiences — a sort of itinerant classroom. To make sure that students understood the purpose of the excursions, the school published a special booklet for each grade. The ninth-grade book, the most extensive of the three at sixty-two pages, cited three wide-ranging objectives:

1. By working together with teachers and each other in an unfamiliar environment — let's develop lifelong memories.
2. By visiting various historical places directly — let's deepen our studies and understanding of our heritage.
3. By working together within a group with good health and safety — let's learn about public manners and have a positive experience.

The book went on to list the places to be visited, the time allotted to each location, and the mode of transportation between sites.

Like the dress code, the guidebook left nothing to chance — it provided rules right down to how to take a communal bath in the hotel's tub. "All students should bathe in shifts," the book insisted. "Fourteen students at a time; ten minutes per group." To ensure that this schedule was followed, the guide indicated that a teacher would be stationed outside the bathroom door every evening, logging students in and out of the tub. Also, like patients being admitted to a hospital, each student was advised to bring a pair of slippers, a towel, and a bar of soap. Shampoo, however, was forbidden, since washing hair would be too time-consuming. In addition, the students were admonished to follow these rules when bathing en masse:

1. Before entering the bath, wash yourself thoroughly.
2. When entering the bath, do not take your towel with you.
3. Upon leaving the bath, do not forget your underwear.

Ironically, these instructions contradicted my own experience with public bathing in Japan, since all of the men in my inaugural bath carried tiny towels into the water with them to shield their dignity. But for students the rules are more strict.

The family bath, or *furo*, is the hallowed yolk of most Japanese homes, a secluded sanctuary where weary servants of the state can wash away the dust of everyday life and renourish their ties with the past. Most parents take their children into the tub with them at night as soon as they are old enough to walk. Many families, like the Cherry Blossoms, drop fresh fruit or flower petals into their bathwater on special occasions throughout the year.

Like eating, dressing, and bowing, bathing is considered so central to Japanese culture that Sakamoto-*sensei* did not trust parents to teach their children the proper form. The school, he felt, must take an active role. I could only imagine the reaction in America if a school administrator set out to teach fifteen-year-old students the "official" way to take a shower. "First take off your clothes. Then turn on the water. Don't forget to wash behind your ears . . ." The secret to Japanese schools, I had come to realize, is their pledge to leave no towel unturned in pursuit of an orderly world.

In the weeks before the start of trip season, students heard in meeting after meeting that their behavior reflected not only on themselves but on their class, their school, and their community as well. Then, two days before the trip to Kyoto, all the ninth-grade classes gathered in the gymnasium for a rigorous uniform inspection. The students stood at strict attention as teachers marched down the lines, scrutinizing the width of collars, the length of hair, and the placement of nametags on jackets and shirts. The teachers stretched rulers to the girls' hemlines and measured the distance from the cuffs of boys' pants to the floor. Any student whose uniform showed deformities was sent to the stage and ordered to strip off the delinquent garment, which was then laid on the floor alongside the prototype that Sakamoto-*sensei* had pulled down from outside his office. Following this operation, the students were asked to hand over their prepackaged luggage so that teachers could search for contraband. No Walkman radios were allowed. Chewing gum was strictly outlawed.

Thinking of the dissension at the sports festival, I pulled some boys aside to ask them which illegal items they planned to stow away.

"I'm bringing popcorn," one whispered, "but please don't tell my teacher."

"I have a tape recorder," said another.

"I'm bringing comic books," said a third.

"No cigarettes or sake?" I asked.

"Are you kidding?" one boy said with a gasp. "Too dangerous. Didn't you hear that one student got sent home by a teacher last year for bringing a video game? Candy is okay, but nothing serious like hair spray or beer."

In the mood of vigilance that prevailed after the suicide in May, Sakamoto-*sensei* balked at the idea of sending me to Kyoto with the ninth graders. Instead, after lengthy deliberations with my office elders, he decided that I could accompany the seventh graders on their trip to Tokyo Disneyland and help chaperon Denver's homeroom class.

On the morning of the big day, the students began assembling at school before six o'clock. The school had tried to anticipate every possible problem, beginning with the coordination of two hundred students and five rented luxury buses. This potential bottleneck was considered so severe that on the day before the trip, the teachers set up chairs in the gymnasium in the shape of a bus to allow students to practice getting in and out of their seats in an orderly fashion. This rehearsal was followed by another, in which students practiced running in and out of formation for the class photograph when they heard the sound of a whistle.

Once on the bus, however, the teachers went off duty and the tour guide took over for the ride. While schoolteachers regularly talk down to their students ("Now, children, let's hold our bags tightly as we board . . ."), the young bus hostess, dressed in a pink polyester uniform with a navy vest and cap, used florid and servile speech to the students as if addressing an honored group of executives ("Good morning, ladies and gentlemen, we are privileged to have you on board our bus today for your trip to Disneyland. Please fasten your seat belts and we'll be on our way . . .").

The students, however, were unimpressed.

"*Sensei, sensei,*" one screamed at Denver from the back of the bus, "can we switch seats?"

"*Sensei, sensei,*" called another, "is she your girlfriend?"

Finally the guide gave in to the pressure and offered the microphone to the class. A buoyant boy with a devious grin sprang from the back of the bus to seize the proffered prize. He whispered something to the guide and handed her a cassette tape, which she popped into a tape recorder at her feet, flooding the bus with the synthesized sounds of Hikaru Genji. Without missing a beat the boy stood atop the purple velour seats of the bus, bowed to the cheers of his rowdy classmates, and began to sing along — "Here I am in par-a-dise." Welcome to junior *kara-oke*, the teenage mutant version of the adult drinking game.

As we neared our destination, the soaring spires of Cinderella's castle peeked over the congested mesh of freeways and factories like a mirage of fantasy in an industrial desert. The guide retrieved the microphone and asked the students to return to their seats. After she reached for a button beside the driver's chair, two doors magically parted above the students' heads, revealing a television set suspended from the ceiling. "Mick-ey, Mick-ey!" the students cheered, and within seconds a tuxedoed Mickey-*san* appeared on the screen to lead us on a Magic Kingdom tour full of song and dance and Disney trivia. "Tokyo Disneyland opened five years ago . . . The park welcomed five million people in its first year . . . We've sold two million T-shirts and three million hot dogs since our opening day . . ." At the end of the tour, Mickey asked everyone to join him in a round of "It's a Small World," the anthem of Disney diplomacy.

Disney. The name itself conjures up images of magic and spectacle, of a world where reality succumbs to fantasy and hope conquers fear. Step inside this rainbow-colored paradise where music fills the air and candy grows on trees and leave your worries behind. Disneyland defies time. It legitimizes escape. In essence, it is the opposite of school.

The teachers freely admitted that Tokyo Disneyland served no educational purpose, but they still viewed the trip as more than fun and games. Chief among their objectives was for students to design a plan for the day with several other classmates in a small group, or *han*. Each *han* was required to stick together for all four hours and incorporate ideas from each of its members, thus ensuring that students experience the actual give-and-take of group decisionmaking. One of the primary rules for the class trip, for example, was that all students be prompt. But the rules insisted that only one student per *han* was allowed to wear a watch. The teachers wanted a diary, so one student in each *han* was elected scribe. The class wanted pictures, so one student per group was allowed to bring a camera. The students wanted souvenirs, so each group collected a kitty and elected a purser to tend it. Dependence on other people is not a genetic trait that the Japanese pass down through the blood, it is the logical outgrowth of deliberate pedagogical policies like these. The management of this trip, more than any other single event in the school year, revealed the skill with which the Japanese schools transfer abstract goals into concrete educational practices.

"Remember our objectives," Denver said as the students gathered for one last time in front of the bus. "Have fun, but do not get lost. Today we want everyone to make good memories. Let's cooperate, and let's be prompt. Be back at the bus by three o'clock."

Tokyo Disneyland is almost identical in style and layout to its cousins in California and Florida. The park has manicured shrubs in the shape of cartoon characters, pink sidewalks that curve around peacock-colored flower gardens, and speakers hidden in all the trees which sing like the birds and the bees. Each area of the park glitters with memorabilia from its chosen theme: Frontierland has wooden storefronts, split-rail fences, and gold sheriff stars etched on every door; Tomorrowland boasts large metallic buildings, UFO-like hot dog stands, and aerodynamic trash cans straight out of a twenty-year-old vision of the future. (Frankly, these futuristic gimmicks pale in comparison to the high-tech gas stations and computerized toilets twenty minutes up the road in downtown Tokyo.)

One major distinction between Tokyo Disneyland and its American forebears is that the Japanese version has no Main Street. Instead, the main shopping strip is called World Bazaar and is covered with a giant glass roof to protect park patrons during busy and drizzly summer months. The lack of Main Street USA, in many ways the epitome of the Disney charm, hints at a larger difference between the two worlds. What is lacking in Tokyo Disneyland is a sense of nostalgia. In the United States, Disneyland aims to divert its guests from the present to a fond, misty memory of some make-believe golden age in America's past. But in Japan, the main street never had soda fountains with large bay windows or bakeries with gingerbread trim. What for Americans represents the romance of the past, for Japanese symbolizes a charmed vision of the future. Instead of a sense of loss, there is a sense of longing.

"Do all streets in America have jazz bands?" one student asked me as we walked by a three-piece ensemble wearing candy-striped jackets and white skimmer hats.

Denver hoped that his students would not only get a taste of sugar-coated Disney but also come away from their excursion with a better sense of foreign culture. With this in mind, he gave each of his students an assignment: find a group of foreigners somewhere in the park, ask them for their impressions of Japan, and secure an autograph as proof of the exchange. When I first became a teacher in Japan, I hated this type of assignment because it seemed to promote the idea that singling out foreign faces on the sidewalks or the trains could somehow promote international exchange. Surely, the dozens of people who had my photograph on their mantelpiece were no less "internationalized" than they were before they ran into me at the laundromat or the supermarket. At school I refused to sign autographs, pleading — sometimes on deaf ears — that I was a teacher, not a celebrity.

After talking with students about these exchanges, however, and hearing Mrs. Negishi's story about speaking English with an American soldier when she was a child, I learned never to underestimate the power of a simple gesture. As a native speaker I was often called upon to judge English-speaking contests in Tochigi. Student after student would rise to the platform in these affairs and relate how a brief hello from a foreigner in Nikko, or a chance meeting with a business associate of their father's, had turned them on to the importance of studying English.

"*Sensei, sensei,*" two girls screamed to Denver later that afternoon in Frontierland, "we met two people from France. We thought they were American, but they weren't. Their English was not much better than ours. They said Japan was beautiful!"

"*Sensei, sensei,*" a group of four boys called as we waited in a one-hour line for a ride on the Space Mountain roller coaster, "a woman from England asked us where we were from. We said Tochigi, but she did not know where that was. Then we told her we lived near Nikko, and she said she had visited there last week."

Denver relished the way his students reacted to this task. "When I was a student in Sano," he told me, "we didn't have such things as American movies and American teachers in Tochigi. When I first went to university in Tokyo, I used to eat at Wendy's every day with my girlfriend because neither one of us had ever been to this kind of place before in our lives. I remember when I met my American professor of English. It took me several months to speak to him, but now he is my friend."

Denver hoped his students would not be as reluctant to assert themselves around foreigners as he had been.

"When you first came to our school my students were very afraid of you. They would come talk to me after class and say they could not understand what you said. They laughed at your long arms and way of speaking. But now they are used to you. They were very happy that you rode on our bus this morning. It made them feel very special."

"Mr. Bruce, Mr. Bruce," said a group of girls who came across Denver and me while we were standing in line to buy ice cream. "We can't find any foreigners anywhere. Do you mind if we ask you a few questions about Japan?"

"Sure," I said, "I'll be glad to answer."

They giggled, and I suspected a trap.

"By the way, will you give us your autograph?"

On the day after the trip to Tokyo Disneyland, Denver convened a special homeroom meeting.

"Good morning, boys and girls," he said at the outset. "I want to thank you all for following the rules yesterday and making our trip a success. I would like all of you to spend the next forty-five minutes thinking about your behavior yesterday and writing a short reflection describing how you felt about the trip. Please begin now."

Like most teachers, Denver regularly conducted such "reflection sessions," or *hansei*, to encourage students to examine their personal conduct and express their feelings toward the class. Like the open discussions after lunch, these exercises were part of the overall effort to encourage a sense of responsibility and allegiance to the *kumi*.

"My group decided to go on one ride in every section of Tokyo Disneyland," wrote one girl, "so we were very busy."

"I met a girl from America," said another, "and I was very surprised that she could speak Japanese. I want to be her pen pal."

The students took this assignment very seriously, and many apologized on their reflection sheets for defying some of the school rules. They offered confessionals of sorts for sneaking candy on board the bus, losing their nametags on a ride, or buying one ice cream cone too many. Although they admitted their violations, the students wanted their teacher to know that they had still learned from him a sense of duty and a desire to care for others.

Dear *Sensei*,

   I had a very nice time in Tokyo Disneyland yesterday. I enjoyed seeing Cinderella's castle and having my picture taken with Mickey Mouse. I was glad it did not rain. But I am afraid that I disobeyed one of your rules. You told us that we could not take more than 1,500 yen per person [about $12]. My group, however, did not have enough money in our collection to buy each person a souvenir. Before I left home, my mother gave me 3,000 yen. I used some of this money so that all of our group members could buy the same T-shirt. Our shirts are very beautiful, and we are very happy today. I hope you are not angry. Now at least we have a pleasant memory.

Sincerely,
Kumiko Tanaka

24

THE AMERICAN CLASS: LESSONS FROM INSIDE THE JAPANESE SCHOOLS

How can the bird that is born for joy,
Sit in a cage and sing.
How can a child when fears annoy,
But droop his tender wing,
And forget his youthful spring.
— *William Blake, "The School Boy," 1789*

"DEAR MR. BLUSE", a seventh-grade boy wrote to me in flawless cursive script at the end of my first week in school:

> I saw for the first time, first at the time, then be surprised at tall.
> Unlike Japanese teacher, English lesson enjoyed.
> For short, but thank you very much for everything.
>
> Love,
> Ryuichi Tominaga

Since my opening day in a Japanese classroom, when I first walked from behind the teacher's podium, offered my hand to an unsuspecting boy, and was met by the blank stares and frightened faces of forty-five fearful students, I was consumed by a singular question: How could I accommodate Japanese classroom manners and still expose my students to an "international" way of schooling?

Ryuichi's reactions were typical. Most students were

shocked at first by the sight of this lanky foreign teacher. They were startled when I jumped around the room, bounded from desk to desk, and asked rapid questions in English. "Do you like Georgia Coffee?" "Does he like going skiing?" "Do they like the Beatles?" Most of the students eventually warmed to this style. Now they giggled when I sat on the teacher's desk or drew funny maps on the board. Sometimes they even answered my questions before I called their names. "Mr. Bruce, I like Georgia Coffee very much."

The teachers were a different story.

Mr. Fuji, for example, had been teaching for thirty-five years, and he did not want a young American coming into his life and encouraging him to be livelier and speak more English in class. He was perfectly content to follow along in the teacher's manual, give vocabulary quizzes once a week, and check memorization every chapter. About the only role he could think of for a native English speaker was to read every sentence in the chapter twice a day and have students repeat it en masse. This was called pronunciation practice: mine improved considerably.

Denver, meanwhile, was the exact opposite. He was so excited to have a native speaker in his class that he would regularly stop by my apartment unannounced the night before we were to teach together to ask if I could think of any more songs or games that we could include in our lesson plan. We would sit for hours on the tatami floor of my apartment devising easy-English versions of such American game show classics as "Wheel of Fortune" and "Jeopardy!"

Answer: "A career woman."

Question: "What does Emi [the quixotic star of the English textbook] want to be in the future?"

Mrs. Negishi, like most teachers, fell in between these two poles. At times she would plan special activities to encourage students to talk with foreigners. Once she hooked up a portable microphone in class, told her students they were reporters, and hosted a mock news conference in which I played the president of the United States: "Mr. President, can you use chopsticks?" She knew that if her students enjoyed these exercises, they would study harder at home. But she also knew that conversation games and sing-alongs were not enough to prepare her students for high school entrance exams, so she plodded along with her thrice-weekly training program of vocabulary drills, diagram workouts, and grammar calisthenics.

In early July, as the time neared for me to leave, Mrs. Negishi approached me with her grandest idea to date.

"Mr. Bruce," she said one afternoon when all the students had gone home, "the students have had a very difficult term. I want to try something different for our final class of the year."

"Okay," I said, "what do you have in mind?"

"I want you to design an American class. You can do whatever you want and say whatever you please, just like in America. I want this to be a treat for my homeroom class."

Of course I readily agreed, and over the next several weeks I prepared what I thought would be a model Team Teaching plan. During the year I had learned that many of my students could memorize a Shakespeare soliloquy or diagram the U.S. Constitution, but very few could talk their way out of a 7-Eleven using simple conversational English. The objective of my "American class," therefore, was to have students spend a majority of their time speaking and listening to English. There would be no recitation of textbook passages; no reading aloud after the teacher; no writing out pronunciation guides using stodgy linguistic symbols. Instead, students would put their knowledge to work and take a chance at "living" English.

One of the first things I insisted for the American class was that students be allowed to wear something other than their school uniforms. "If this were really America," I told Mrs. Negishi, "they could make their own choices. Class will be more colorful this way, and students will have more fun." But as soon as I stepped into the ninth-grade classroom on that hazy, humid July afternoon, I feared that my dream had gone awry. Instead of vivid greens and neon oranges, all the boys stood in a line in front of the blackboard dressed in virtually identical outfits of blue jeans and white T-shirts. The girls, on the other side of the room, showed a little more fashion courage, but most wore denim skirts and baggy tops with polka dots or stars. The students looked like nervous high school freshmen at their first sock hop, and the room, which had been decorated with stenciled drawings of Donald Duck and Betty Boop, looked like the local headquarters for the Mickey Mouse Club of Tochigi. "Welcome to America!" a student had chalked on the blackboard. "A friend in need is a friend indeed."

My second proposal for the American class was to move the students' desks from their normal rigid structure and set them in a circle. It seemed ironic that I was preaching the virtues of circular seating in Japan, but Mrs. Negishi said she had never seen it done before. My purpose was to open the fixed desk alignment so that all students had an equal chance — and an equal risk — of speaking in class. But once I moved in front of the blackboard, I realized why no one had tried this technique before. With forty-five students in one class, the desks would not fit into a single loop, so the students had to cram them into two rings that looked less like a fine-tuned platoon and more like a pileup of bumper cars. Undaunted, Mrs. Negishi and I stepped into the center of the room and asked the students to stand up. Instead of asking them to bow, however, we sprung a trap: no student would be allowed to sit down until he or she had correctly answered a question in English.

On your mark, get set, go.

We darted around the room, turning first to one student ("What did you eat for breakfast this morning?") and then to another ("What did you watch on TV last night?"). If the answer perchance was correct ("I watched Best Hits USA"), the student would cheer and slide into a chair, but if the verb was missing or the object misplaced, the student had to remain standing until we came around again. With persistence and some coaching, all the students eventually talked their way into their seats.

Flush with success from our surprising opening, I moved to the blackboard for a quick review of the material in the lesson. Instead of reading the six pages on Singapore in the textbook and having the class repeat after me, I gave the students a short tour d'horizon with my chalk, stopping off at the United States, China, Hong Kong, Thailand, and Malaysia before arriving at Singapore. Even with the textbooks closed, we covered all the material in the chapter ("Singapore is a clean and green country"), introduced the vocabulary words ("You must pay a fine if you *litter* in the streets"), and even reviewed the "target sentence" ("*Have you ever* . . . been to Singapore?").

Picking up on this target theme, I asked the students to open their notebooks and gave them one minute to write a sample question using the sentence "Have you ever . . . ?"

"Mr. Bruce," asked the first girl I chose, "have you ever been to American Disneyland?"

"Yes," I said, "I have been there three times."

"Have you ever eaten sushi?" queried the next.

"Yes, I have eaten it many times."

"Have you ever eaten breakfast before five o'clock?"

"No," I replied, "I don't believe I have."

Following this game, Mrs. Negishi and I distributed a model dialogue to help students practice what they had learned. With twenty minutes left to go in the class, we divided the students into groups of two and asked them to fill in the blanks of our dialogue. As they worked, we rearranged the chairs and focused Mrs. Negishi's video camera on the makeshift stage in the center of the room.

"Are you ready?" I called, plucking two volunteers from the

circle and thrusting them into the middle. "Take one," I said, mimicking the director's routine I had learned at the *juku*. "Begin!"

"H'lo," the first boy mumbled, with his eyes glued to the handout and his mouth hidden behind the paper.

"Hunh?" came the reply.

"No, no, no," I interrupted, waving my hands in exaggerated enthusiasm and snatching the paper from their hands. "Bigger, faster, louder. Be happy!"

"Hello," the two boys screamed at once.

"My name is Masatoshi."

"My name is Yuji."

"Nice to meet you."

The two boys shook hands hurriedly, then rubbed their palms on their jeans.

"Have you ever been to Utsunomiya?"

"Yes, I have. I have been there many times. Have you ever been to Alaska?"

"No, I have not."

"That's too bad. I hope you go there someday."

"Thank you very much."

"You're welcome."

"Good-bye."

After months of hearing nothing but nonsensical exchanges, this dialogue was poetry to my ears. During a year of teaching English in Sano, I had learned to measure progress not by lists of words or flights of rhetoric, but by the simple magic of call-and-response. "Speaking English is like tennis," I often said. "When you receive the ball from your counterpart, you have to hit it back. If not, the game cannot continue." From my earliest days as a teacher, I preached the gospel of speaking up and talking out. "Speak up when a question is asked," I encouraged. "Talk out when you are in class. Whatever you do, just let your voice be heard."

After several more students took their turn in the central

ring of Room 9-2, Mrs. Negishi stepped in to say that class was almost over and we would have time for only one more dialogue. Six hands shot into the air. She walked slowly around the perimeter, eyeing each of the candidates, and then finally selected a slim boy with glasses who was the star pupil of the class.

"Susumu," she said, "you will have a conversation with Mr. Bruce."

The class applauded. The boy rose cautiously, ran his hand across his crew cut, and took his place inside the circle.

"Good afternoon," I said. "My name is Bruce."

"Good afternoon," he replied. "My name is Susumu. Nice to meet you."

"Nice to meet you."

We shook hands.

"Have you ever been to America?" the boy asked.

Several of the girls started giggling.

"Yes," I said, "I have been there many times. Have *you* ever been to America?"

"No, I have not."

"That's too bad. I hope you go there someday."

"I hope so, too," he said. "I want to visit Georgia."

I glanced at my mimeographed sheet but suddenly felt at a loss for words. "I see."

"When I drink Georgia Coffee," the boy continued, "I always think of you. So I want to visit Georgia. Then I want to listen to your loud voice. I love it."

"Thank you very much," I said, offering my hand as I had taught them to do.

Susumu took my hand and started to shake it, but changed his mind and bowed instead. "Thank *you*, Mr. Bruce," he said. "I am happy today."

"Me too," I said, bowing back. "Good-bye."

"Good-bye."

·   ·   ·

I began my year as a teacher in Tochigi seeking clues to Japan from inside its schools. Many of the country's strengths, I discovered, are passed on early to students: discipline is taught through the strict dress code, cooperation through the *kumi*, and responsibility through the give-and-take of the teacher-student bond. But Japan's shortcomings are born here as well: stress is nurtured through "examination hell," pressure through the threat of gang bullying, and intolerance through the lingering myth that the Japanese are a breed apart. As an American in the Japanese schools, I found myself torn between respect and aversion for this system. On one hand, I admired students' good study habits and strong communal values, but on the other, I was concerned about the harsh rules and do-or-die exams that schools use to achieve these results.

In an effort to make sense of my conflicting feelings, I searched for a new way to evaluate schools, one that would look beyond the standardized test scores and consider more than the "Three R's" of reading, writing, and arithmetic. What I settled on was a broader standard, based on what I call the "Four C's": curriculum, communication, character, and citizenship.

Curriculum. One of the undisputed strengths of the Japanese school system is its ability to teach children cognitive skills, particularly in math and science. All public schools in Japan follow a curriculum established by the Ministry of Education in Tokyo, ensuring that all students are taught the same information at roughly the same time of year. While the Western philosophy of education is based primarily on the dialogue, in which the teacher and the students exchange information, the Japanese system is based on the monologue, in which the teacher speaks and the students receive. This style, with its stress on lectures and rote memorization, is particularly suited to teaching math and science skills, especially at an early age. Every major international study of the last fifteen years has shown that Japanese children consistently outperform their Western counterparts in these two areas.

But while the Japanese clearly excel in teaching cognitive skills, they lag far behind in teaching creative thinking. The same monolithic teaching methods that work wonders in teaching mathematical formulas and scientific data are less successful in encouraging children to interpret historical trends and express themselves in a foreign language. A Japanese friend who studied with an American university professor in Osaka once asked me to help him with an English essay on the causes of the Second World War. My friend had written a three-page essay outlining the opinions of eminent historians on various sides of the issue. His analysis was very thorough, but at the end he stopped short of weighing the assorted ideas and stating his own opinion.

"Which side do you agree with?" I asked.

"I don't know," he replied. "I didn't think about it."

"The professor will be interested to know your view on this issue. He wants you to express your opinion."

"But I am just a student," he said. "I am not supposed to have an opinion."

Of course Japanese children *can* be creative, but in school they are taught to bring their thinking into line with that of others, rather than to draw conclusions for themselves. In art classes at Sano Junior High, for example, the teacher distributed two dozen busts of Michelangelo's David and told each student to make an exact, frontal drawing of the face. No deviations were allowed. Many Japanese are now starting to question the lack of originality in their schools. Taichi Sakaiya, a well-known Japanese writer and former official with the Ministry of International Trade and Industry (MITI), stressed recently that Japan's success as a world leader of mass-produced goods is the result of the country's standardized education and the technical skills of its workers. "But individuality and variety have been sacrificed for this goal," he wrote in a newspaper editorial, "and as a result, contemporary Japan, though materially abundant, is a most uninteresting place." His proposed solution was to shift the focus in schools from serving the needs of society toward fostering independence of mind. "The concept of standardization must be discarded," he insisted, "and society refashioned so that individuality and creativity have a place. The continued prosperity of Japan depends on whether social change can be moved in this direction, because future success in business will require innovation."

Communication. After my year in the Japanese schools, I believe that the largest gulf between Japan and the West is caused by our dissimilar ways of communicating with others. Westerners often ask how I can possibly *understand* the Japanese: They are so quiet. Their faces are like stone. They never show any emotion. The Japanese, for their part, say similar things about Westerners, particularly Americans: They talk so much. They always say whatever they think, even if it is rude. They are inconsiderate of other people's feelings.

I came to appreciate that Japanese reticence and indirect speech are not the result of a lack of emotion or a willfulness to deceive but are the outcome of careful and deliberate training. Students are constantly reminded not to interject their personal feelings into a public discussion but to save them for a more private time. Among themselves, the Japanese are masters of the art of not offending anyone. To them, this indirectness seems considerate and politic, while to us it seems evasive and, at times, maddening.

My biggest challenge as a teacher in Japan was to emphasize that while this style is effective among Japanese, it is less effective when speaking with people who are not Japanese. "English is more than a language," I would often say to my classes, "it is a state of mind. If you are speaking English the same way you are speaking Japanese, then you are probably speaking it incorrectly." I first realized this different approach through my own experience with the Japanese language. When I spoke Japanese with Mr. C, my eighty-five-year-old landlady, or the principal of my school, I could feel myself physically change. I would scrunch my shoulders, stiffen my arms, and even suck in my stomach. At times my posture would dictate my speech. It's nearly impossible, for example, to have a good knock-down, drag-out fight in standard Japanese. One's body — one's words — will not allow it.

English, on the other hand, is perfect for confrontation. Not only does the language favor frankness, but children are taught to be as direct and succinct as possible. Even our idioms are different: Japanese are constantly reminded to be polite, considerate, and mannerly, while Americans are told to "talk straight," "get to the point," and "put it on the line."

The value of learning a foreign language is to understand these basic distinctions. All Japanese students beginning in the seventh grade must study English for three years, plus three more if they attend high school. This represents a commitment to language study which American schools would do well to follow. But the obstacle in many English-language classrooms in Japan is that very few teachers emphasize cross-cultural themes. Many Japanese instructors teach English as if it were Japanese: through constant memorization and repetition of words. Every student at Sano Junior High had drawers full of old notebooks with page after page of English vocabulary words written ad infinitum. A student might be able to learn the Japanese characters for "*sensei*" by copying them fifty times, but writing the English word "teacher" one hundred times a night does not guarantee that a student will be able to understand it, or use it. An important component of communication is conversation, a device that few teachers like to use because it is not covered on the entrance exams. If schools hope to breed "international" students, then Japanese teachers must adopt techniques that stress speaking and listening skills as basic steps toward interpersonal communication.

Character. For all the rules and regulations that govern their lives, Japanese students still enjoy school. When asked to describe school life in my year-end survey, over two thirds of the students wrote such phrases as "cheerful," "comfortable," "noisy," or as one student aptly expressed, "spirited, at least between classes." But the spirit that lives in the halls between classes often dies out when students return to their seats.

During my time in Japan I met a sixteen-year-old girl who had spent seven months in Arizona with her family while her father worked at an American computer company. The girl, named Mia, attended eighth grade in a Phoenix-area junior high school and was struck by the different behavior of American and Japanese students. "At first I was shocked by how many students raised their hands in class," she said. "They all wanted to speak. They all wanted to answer the questions. They all wanted to talk with the teacher. That would never happen in Japan."

In the beginning of her sojourn, Mia hated her American school. "I want to go home," she cried to her parents. Her math class was too easy, and her history class was too hard. The students were noisy in class, and she missed the familiar structure of school life in Japan. But at the end of the year, when Mia returned to Japan, she experienced a similar shock in reverse.

"Japanese classes seemed so boring," she remembered. "Nobody raised their hands. Nobody answered questions. Nobody spoke in class except the teacher. School was like a factory pouring information into students as if we were all canned peaches."

Mia is not alone in her frustration. Even the Ministry of Education has entered this debate and is now calling for more focus on the character of individuals in Japanese schools. In a blueprint for educational reform prepared for the prime minister in the late 1980s, the ministry appealed for a new age of individuality: "Educators in Japan should once again clearly grasp what 'perfection of human character' means, and put emphasis on the importance of personal dignity and respect for individuality, which, in the process of Japan's rapid modernization to catch up with the West, tended to be ignored."

As its minimal dropout rate can attest, Japan is extremely

successful in convincing children that their future success depends on staying within the system. But if schools hope to maintain this level of commitment, they must spend more time cultivating in students a sense of personal fulfillment.

Citizenship. One of the most impressive aspects of the Japanese school system is its ability to foster among students an allegiance to the state. In the United States, students learn early about their rights as Americans. "It's a free country," adolescents often say, "I can do whatever I want." But do they learn about their responsibilities? In Japan, this balance is made clear. From cleaning the windows of their classroom to picking up trash in their neighborhood, students learn the importance of serving their community. The essence of citizenship, I believe, is the feeling that people value their place in a group to such an extent that they are willing to sacrifice some of themselves so that the group as a whole may prosper. Japanese schools' biggest service to the state is their ability to create among most students this overwhelming sense of belonging.

Yet for all their paeans to group cooperation, Japanese schools still fall horribly short in teaching children how to get along with those who look, think, and act differently from the majority. The beleaguered minority in schools includes handicapped students who are assigned to special classes, "returnee" students who have lived abroad, and students who are descended from families that were outcasts over a century ago. This intolerance is no more apparent than in the way young Japanese feel toward foreigners. When I asked students in my survey if they thought Japan was superior to other countries, over seventy-five percent said yes. Their reasons: the Japanese are more honest than other people; they work harder; and, the most popular answer of all, the Japanese have better brains. This fact, more than any other, raises doubts about the ability of young Japanese to live in an international world. While Japanese schools prepare their students to be citizens of Japan, they fail to teach them to be citizens of the world. If these students can learn one thing from the West, it is respect for diversity. The government's Ad Hoc Council on Reform issued this warning best: "Students are expected to love Japan as Japanese, but they must avoid judging things on the basis of narrow nationalistic interests."

Japan, like its economy, cannot exist in a vacuum purposefully cut off from the rest of the world. More than ever before, the future success of the Japanese people is linked to the continued good fortune and good will of people living across Asia, the Americas, and Europe — all former enemies of imperial Japan. Thus, it is in the best interest of Japan — and of other nations as well — for Japanese schools to raise a new generation of students who will be able to blend with people all over the world, even those not "Made in Japan."

The last chapter of the ninth-grade moral education textbook at Sano Junior High addressed this issue through the story of a young junior high school student named Masao. At the beginning of the tale, Masao receives a grade on a midterm exam that says he is third in his class.

"It's not bad," he thinks, "but I'll try to become the top student. I just need to study more than anyone else."

For the next several months Masao spends almost all of his time studying. He never watches television; he rarely leaves his house; he doesn't even speak with his best friend, Takeo, the number-one student in the class.

Six months later, when the results of the final exam are made known, Masao has moved to the top of the class. His hard work has paid off, he thinks. He is satisfied and feels superior. But when the scores are announced before the class, a chill fills the room. The other students do not applaud the top scorer as they usually do. Masao is shocked, but he assumes the other students are envious of him.

Several weeks after the test, Masao's old friend Takeo approaches him and asks if he would like to help produce the class yearbook. Masao thinks Takeo is trying to take away his studying time and refuses the request. The next day Masao's teacher asks the boy to stay after school for a talk.

"Congratulations," the teacher says. "You did very well on the last exam, the top score in the ninth grade."

Masao listens quietly.

"By the way," the teacher says, "do you know the phrase 'Made in Japan'?"

"It means something is produced in Japan," the boy says.

"There is more to it," the teacher says. "Japan has improved considerably in the past twenty to thirty years. We now have the world's third-largest GNP. But how much do other nations trust and admire us?"

Masao is confused but listens anyway.

"Other nations think that Japanese people study and work hard," the teacher continues. "This has certainly helped Japan to grow. We think we can be proud and others should admire the effort we have showed. But they do not. Some are envious of us, but I am afraid that most think Japan does not do enough to help others, especially countries less fortunate than ours."

Slowly the boy begins to understand.

"We are very wealthy, at least in material goods. But can we ignore the sad reality of others because it is not our problem? I think Japan ought to think of its position and try to help other nations. Then 'Made in Japan' will be accepted as the mark of a good product."

As he listens, the boy at last understands the reason for the chilled air the day the scores were announced. Just as Japan needs to help other countries, so he should contribute more to his homeroom class.

"I was the one who suggested you work on the yearbook," the teacher says. "I am disappointed that you turned it down. Why don't you think it over?"

As Masao walks away, he thinks of asking his teacher why

he originally spoke so indirectly. Instead, he nods and smiles. He has learned what it means to be "Made in Japan": those who are strongest individually must work even harder to preserve the health of the larger — and stronger — fold.

About a week after the American class, I was sitting at my desk at the Board of Education late one afternoon when Sakamoto-*sensei* appeared at the door. He had never visited our office before. After pausing at the door to announce his name, he stopped briefly by my desk on his way to meet the director.

"Bruce-*sensei*," he whispered in my ear, "thank you for teaching our students in such a kind and gentle way last week. They love you from the bottom of their hearts . . .

"By the way, if it's all right with you, I would like you to give a short speech to the entire school next week at our summer farewell ceremony. I think the students would enjoy hearing from an American teacher. These are extraordinary times."

I looked up to thank him for the invitation, but he had already moved on to greet Kato-*sensei*, who was waiting beside his desk. The two men then walked slowly to the back of the room and disappeared into a private chamber.

I looked at Mr. C, who whispered something into the telephone and quietly hung up the receiver. He pulled a cigarette from a pack in his drawer and let it dangle from his lips without lighting it. I returned to the work on my desk.

After several minutes of silence, the two men emerged from behind the door and stopped just in front of the director's desk.

"Excuse me, ladies and gentlemen," Kato-*sensei* began, "I trust that you all know Mr. Principal. I believe he has a short announcement to make."

Sakamoto-*sensei* stepped forward and dropped his arms by his side as he did before addressing his school. His shoulders were straight, his face polished, his eyes locked in a timeless stare as in the tinted pictures of past principals which lined the walls of his office.

"As you know, ladies and gentlemen, these last few months have been very difficult . . ." His voice trailed off slightly but then regained its gravelly tone. "A sickness has been in the air, a very serious malady. I am afraid that this illness has finally infected me."

None of the members of my office moved as the principal spoke, using what had become a familiar metaphor to describe what had happened at my school. I recalled Denver's warning that after a disgrace to a child, a lord must bear the shame.

"My colleagues and esteemed members of the Board of Education, I am announcing to you today that I will be retiring from my job at the end of this term to recuperate from this disease. I apologize for any inconvenience this may cause. Thank you for your service to my school."

When he had finished speaking, a cool draft seemed to spread through the room like contagion in a bath. Mr. C diverted his eyes to the floor. Arai-*san* drew her hand to her lips as if to repress a cry. Kato-*sensei* stared out the window.

Finally, with his head still bowed, Sakamoto-*sensei* walked out of the office and disappeared through the elevator doors. This "one-man," who had expressed such personal pride in the storied tradition of education in Japan, who had believed so deeply in the need for teachers to instill in children a sense of discipline, who had striven so diligently to teach his students about opening and closing the *kejime* doors of responsibility, in the end had borne the burden of history himself so that his school could persevere.

He would not return that fall.

The following Tuesday morning, I stood in the hushed gymnasium and prepared to take my turn before the entire school. A year had passed since I first stood fresh from my bath before a roomful of teachers in my Lilliputian cotton kimono with the dark blue bamboo print and answered questions about my resilience to Japanese wine, women, and thunder. During that time I had learned how to deflect unwanted questions from eager

Japanese interrogators, how to blend with the seasonal winds, and how to charm an audience with the subtleties of indirect speech. This was my opportunity to speak to my students in their own metaphor.

When my name was called, I walked to the center of the darkened stage, now stripped of its *hinomaru* sun, bowed to the teachers along the wall and then to the students in rows on the floor, and began speaking in my best Japanese.

"It's summer," I said. "In summer, I like to take walks.

"The other day I was walking through Sano when I came upon a doll store on a small side street. From the window, all the dolls in the store looked the same. They had the same costumes, the same posture, the same faces. But I decided to go inside.

"Once inside the store, as I watched the shop master work with his dolls, I realized that they were not all the same. Each doll was different. Their costumes were different; their faces were different. Each had its own character and story to tell.

"In the past year, I have visited many schools. From the outside, all the students in these schools looked the same. Their uniforms were the same, their bicycles were the same; their haircuts were the same. But I went inside and talked with many of you. I found that all of you are not the same. Each of you has a different character and a different story to tell.

"Next year you will have another foreign teacher in your school. At first this person will look like all other foreigners to you. But please don't point at him, and please don't stare. Instead, talk with this teacher, as you talked with me. Ask questions and try to discover his character and the stories he has to share.

"This, I believe, is the best way for different people to understand each other. This way we can all be friends.

"Thank you very much."

The students were silent when I had finished my speech. I stepped back from the microphone, bowed, and turned to leave the stage. As I moved toward the stairs, a young girl came hurrying onto the platform with a bouquet of yellow roses. She laid the flowers in my arms and unfolded a piece of paper from her pocket.

"Dear Mr. Bruce," she began in soft, clear English. "You have visited our school for a year, but now you must go home. We are sad to be parted.

"When you came here first, you shook hands with all of us and said, 'Nice to meet you.' We can never forget that. And I'm sure it will become a nice memory for us. Thank you for visiting our school, and please come back to Japan.

"Finally, let's part without saying 'Good-bye.' I would rather say 'See you again . . .' Well, see you again."

EPILOGUE:

IN SEARCH OF
THE JAPANESE DREAM

As I turn and look at the Plain of Heaven,
The light of the coursing sun is hidden behind it,
The shining moon's rays can't be seen,
White clouds can't move, blocked,
And, regardless of time,
We'll tell, we'll go on talking
About Fuji, this lofty peak.
    — *Yamabe Akahito, "Looking on Mt. Fuji," c. 744*

MY AMERICAN FRIENDS wanted to take Kentucky Fried Chicken; my Japanese friends wanted to take rice balls. We took both, plus a bottle of champagne, and headed for the top of Japan.

It was a hot and humid summer Sunday evening. August had come. The rains had stopped. My term had ended at the Board of Education. But before I moved away from my wind-swept home, I persuaded my friends Ben and Emmett, visitors from Georgia, and Cho and Hara to join me for an all-night trek to the peak of Mount Fuji — the tallest mountain in Japan and the symbolic heart of this symbol-sensitive land. In a country where the flag still causes riots and the national anthem is rarely played, Mount Fuji, with its flawless, silent silhouette, conical features, and razor-sharp ledges, has become a welcome substitute as a national totem. Fuji — the *kanji* characters mean prosperous man — is such a popular emblem that it has been used to name everything from TV stations to banks, from restaurants to photographic film. In every logo the image is the same: two sheer sides rising symmetrically from the ground, topped by a flat, volcanic plateau, and crowned with a cap of white snow that drips down the sides like vanilla icing on a storybook birthday cake.

For over ten months of the year, Mount Fuji is draped in snow and ice and is virtually impassable. But from the middle of July to the end of August, the snow melts, the bald head of the mountain emerges, and climbers are allowed on the trails. Because it is such a cultural landmark, over three hundred thousand people make their way to the top of the mountain every year during the six-week hiking season. As a result, climbing Mount Fuji, like taking a bath, is a communal experience. "You're a fool if you don't climb it once," the Japanese say. "But you're a fool if you climb it *more* than once."

At seven-thirty in the evening we loaded our fried chicken and rice balls into backpacks, tied raincoats and sweaters around our waists, and boarded a bus that would take us from the heart of Tokyo to a designated starting point halfway up the mountain. The trail to the top of Fuji is divided into ten stations, and most climbers start at the fifth level around ten P.M. and hike through the night in order to arrive at the two-mile-high summit (3,776 meters) just before dawn the next day.

At the fifth-station gift shop each of us purchased a plain walking stick with a rising-sun flag and a ribbon of bells attached to the upper end. Before starting, Hara insisted that we take off our packs, remove our rain gear, set down our flashlights, and follow him through a routine of warm-up calisthenics like the ones he performed at his office every day. Just as we started our stretching and chanting an American woman came rushing over to our group.

"Excuse me," she said, "do you know which direction to go?"

"No," Cho said. "Up, I think."

"Oh, I'm sorry," she said. "You just looked so professional, I thought you knew what you were doing."

We eventually found the trail ourselves and set off up the mountain. At first the path was wide and the hikers sparse. The moon was full, but it faded at times behind sullen clouds that threatened from above. In the distance the orange lights of Tokyo flickered in the night, hurried and frantic like a pinball machine. As we walked, we kicked stones along the path, slapped bugs on our arms, and traded blame for agreeing to this trip. We talked about Hawaii, where Hara had just been on his honeymoon and where Ben and Emmett would have preferred to have been that night. Still we marched, like lemmings, heavenward.

After little more than an hour, we passed through the sixth station, a small resting zone with a cluster of wooden cabins where several old women peddled warm milk in bottles, dried squid in bags, and spots along their futon-laden floor for travelers too tired to continue. Out front, two young men with bandannas on their heads and tattoos on their arms sat before an open fire and charged one hundred yen (about sixty-five cents) to burn hot-iron stamps into the walking sticks of suckers like us. The stamp showed that we were 2,400 meters above sea level.

After the sixth station the path grew gradually thinner and the crowds suddenly thicker. In no time we were surrounded by families with young children, small bands of businessmen, and the ubiquitous octogenarian tour group.

"The Japan Travel Association Mount Fuji Hiking Group will now begin walking toward the next station," boomed a tiny lady with a pink cowboy hat and a baby-blue bullhorn. "Let's make sure everyone is together and not separated from our group. Please count off . . ."

Some of the hikers seemed to have spent weeks preparing for the expedition. Most of the men over fifty, for example, came dressed in the same outfit: a red plaid shirt, dark wool knickers, and an olive-green Robin Hood hat. They looked like miniature porcelain dolls from a Swiss chalet gift shop. Hiking gear was not required. Most of the younger men wore sneakers, and I saw more than one woman in heels. Yet everyone, absolutely everyone, carried a stick. These sticks were not actually necessary. The path was fairly smooth from the constant tread of feet, and the crowds had grown so dense that we had no place to fall. The reason we all had sticks was that we had fallen for a scam. The fifth-level shop had sold the sticks dirt cheap, and the sixth-level stop had offered stamps at cut rate. But the higher up the mountain, the more elaborate the stamp, until, by the eighth and ninth stops on the trail, each new seal was priced at twice the rate of the last. Hara — the businessman — cursed himself for not thinking of this idea first; the rest of us — the pushovers — waited dutifully in lines for our Fuji proofs of purchase: 2,700 meters; 3,100 meters.

By early morning the stations seemed to be getting farther apart and the bites of chocolate less effective in stemming the waves of fatigue. As we made our way up the mountain, we could always tell when we neared a station, because each one had its own bouquet of septic air. With ten thousand people climbing every night and only several makeshift latrines at each level, the stations seemed to loom over every horizon, disabling those below with the burden of their stench. By the ninth station — 3,400 meters — the smiling faces had given way to scowls, the lines at the outhouses had become longer, and the piles in the sleeping cabins had grown larger. The only group of climbers who seemed not to tire were the children. One class of elementary students marched up the mountain in shorts; a junior high group wore bicycle helmets to protect against falling rocks; and at least one club of high school students carried bicycles to the top for a giddy ride down the back side of the mountain, like freedom-seeking graduates after their rigorous climb through school.

"Hey, Cho," Ben called when we stopped for a rest in between stations, "would you like to try some chicken?"

"No, thank you," he said. "I prefer Japanese rice."

"You know we eat rice in Georgia," Emmett said. "Except ours doesn't turn sticky like yours, because we put butter on top."

"Sounds good," Cho said, lying.

After five hours of straight climbing, our bodies were drunk with exhaustion. We stopped speaking. It began to rain. Now far above the tree line, I stopped to take a time-exposure photograph of the mountainside at night. Up above, blinking flashlight beams wove their way toward the quilt of clouds, and the whimpering bells attached to every stick filled the air with their plaintive song. The scene was both enchanting and eerie, like a massive funeral dirge. With the sleepy foot soldiers, the stench, the stations, and the thickening mist, I thought of Dante's famous tour and wondered whither this path would lead.

Twenty minutes later, after we passed through the cover of clouds, the top of the mountain came into sight, and with it a solid mass of people clogging the trail between us and it. There we were, 3,600 meters above sea level, the ninth station behind us, three layers of winter clothing on our backs, trapped in the absurdity of a predawn traffic jam. As far as the eye could see, people filled the path, and the trail looked more like a stadium ramp after a football game than the face of a mountain at four o'clock in the morning.

"Come on," Ben shouted, "let's be American."

Determined not to have come all this way only to see the sun come up from the *side* of Mount Fuji, we sneaked along the edges of the path, tiptoed over the slouching bodies, and even — in a fit of nonconformity — stepped outside the ropes, to the gasps of the crowd. It was a race, of sorts, against the oldest clock in the world.

Finally, at 4:55 A.M., dragging our soggy chicken and sodden rice balls, sprinting to the end, we reached the top of the mountain. Never mind that we ran smack into a virtual mall of souvenir shops; never mind that there was no place to sit down without buying a can of coffee or a cup of tea; and never mind that the final branding station wanted two thousand yen (about fifteen dollars) for the ultimate Fuji stamp. We overlooked all of this, plus the smell, the crowds, and the freezing rain, because of what we saw. Standing on an isolated spot atop this ancient volcano, far removed from the crowds of Tokyo and the clogged slope of what surely must have been the most congested mountain in the world at the time, at several minutes after five o'clock on an early-August Monday dawn, looking down on other mountains 'round on all four sides, we popped open our bottle of imported champagne and beheld the sight that has inspired a nation since the gods of heaven first descended the bridge to the Land of the Central Plain: the caress of the Rising Sun.

Later that evening, as we sat around Hara and Emiko's apartment in Tokyo, soaking our worn feet, drying our wet clothes, and eating Domino's pizza, I called back to Sano to make arrangements to drop by the Board of Education to say good-bye. First I called Mr. C's house, but no one answered. Next I telephoned Kato-*sensei* and learned the reason why: Mr. C's mother, who had struggled with an illness and had been in and out of the hospital all year, had died over the weekend. The funeral had been held that morning, Kato-*sensei* said, but perhaps I could pay Mr. C a visit the next day.

On Tuesday morning, with Ben and Emmett content to wander around Tokyo seeking Hard Rock Café T-shirts and Dunkin' doughnuts, Cho and I headed back to Tochigi.

"It seems I have spent most of the last year in the passenger seat of your car," I commented as we headed out on the Tohoku Expressway back toward the countryside.

"NIKKO — 80 KILOMETERS," a sign said.

"Have you ever been to Nikko?" Cho asked, and we both laughed. The question had come to symbolize not only the precarious joys of teaching "Living English" but also the mixed blessing of living in Tochigi, where the best, and perhaps only, place to visit is this famed national park. Cho had been there five times in the previous year; I had been four.

We had managed to find other things to do besides traipsing around Nikko every season. We had gone hiking, and swimming, and even golfing — in the snow. In June he had driven me to the prefectural museum in Utsunomiya on a Sunday afternoon so that I could see the four slivers of bone that constitute the "Oldest Man in Japan." After he broke up with his girlfriend, we had talked, plotted, and planned, but we had actually never gotten around to trying *nanpa*. Cho had even had his hair permed to improve his chances. (This hair style was so popular among young Japanese men that one week later Denver appeared at my apartment to show me *his* new perm. His hair was so wavy, in fact, that he no longer looked like John Denver.)

"I've decided," Cho said as we neared Sano that afternoon.

"Decided what?" I said, remembering his abrupt announcement that he had broken up with Chieko.

"I've decided that I'm going to take the test to become an overseas teacher for the Ministry of Education."

"That sounds great," I said.

"I think it is important for people to live in a foreign country," he said. "I think it is important to understand foreign people. If I go, I can become a better teacher and make many friends. What do you think?"

"I think it's a wonderful idea," I said. "Where will you go?"

"Anywhere," he said. "But they usually send female teachers to Europe or America. So I will probably go to the Third World. I think that sounds more interesting."

As we pulled into town, Cho stopped his car at a stationery shop just blocks away from Mr. C's home.

"We must each give a small remembrance when we arrive at the house," he said. "Not very much is required — several thousand yen will be fine. But we must put it in the proper envelope."

Inside the store I faced the familiar bank of envelopes, with their various ribbons and decorations sprouting from every fold. We picked a plain white model with a black ribbon knotted solemnly in the front, traded our old bills for new ones at the cash register, and scripted our names on the front. Now we were ready to pay our call.

Mrs. C ushered us into the living room, where her husband was seated on the tatami floor with five or six other men from the office. When we entered, the group rose silently and bowed deeply, first to us and then to our host. Mr. C stepped toward the door and led me by the arm to the front of a shrine that took up most of one wall in the room. Huge pine branches stood like sentinels on two sides, flanking a three-tiered wooden structure draped in black and white fabric, covered with bowls of fruit and nuts, and topped with a large framed photograph of Mr. C's mother.

Following Cho's lead, I walked to the front of the elaborate memorial, bowed, laid my envelope in a brass bowl on the floor, and clapped my hands twice.

"Thank you very much for coming," Mr. C said as I returned to the center of the room and took my place around a table with the other men. "You are very polite. Your bowing is very beautiful. You even brought an envelope . . ."

"I was very sorry to hear of your loss," I said, using the formal expression that Cho had taught me in the car.

Mr. C paused to acknowledge my comment, then continued his previous thought. "You have become just like the rest of us," he said. "You talk like us. You sit like us. Even your face now looks like ours."

The others leaned forward to examine my appearance and nodded in consent.

"I agree," said one of the men, who was still staring at me. "His face *has* changed in the last year."

"His eyes are more narrow," added another.

"His nose is less high."

"Yes, I think you have become Japanese," Mr. C concluded. "But there is one thing you must know."

"What is that?" I asked.

"You should never clap out loud at a funeral. Just bring your hands together, but never make a noise."

"I'm very sorry," I said.

"Don't worry," he responded. "It's a simple mistake. Cho-*sensei* did the same thing."

The others laughed and clapped their hands together, silently.

"Anyway, we are glad you came," he said. "Now, we drink."

Mrs. C appeared with a plate full of teacups and lemon sponge cakes and set them down on the table. The mood lightened as we passed around the refreshments. The other teachers did not dwell on Mr. C's mother, except to explain that this indoor memorial was a Shinto tradition. Although most Japanese have Buddhist burials, they said, Mr. C chose to have Shinto rites because of his family's close connection to the neighborhood shrine.

Soon the conversation returned to my remarkable transformation.

"The other day I took Mr. Bruce to the doll shop to buy his farewell present," Kato-*sensei* said. "He looked at each doll very carefully and asked the master how it was made. He was very curious and polite. I was surprised to find many Japanese aspects in his attitude. He gave me the impression of being *more* Japanese than a Japanese.

"But," he continued, now turning to face me, "over the last year I have also been trying to understand your country, the United States, through you. I found in you the energetic power of America, and also the Frontier Spirit. You came to our country and lived by yourself. You taught with great enthusiasm, using your hands and your face. You even use your hands when you talk on the telephone. This is very interesting for us. You have been an excellent diplomat for your country."

The others raised their glasses and drank a toast of tea on my behalf. "The Frontier Spirit," they repeated to themselves, ". . . more Japanese than a Japanese."

"Thank you for your comments," I said. "You are very kind, and I have learned a lot from you." I paused briefly and took a sip of tea. "But I would like to remind you that I am not Japanese. I am an American. I look, think, and act like an American. Sometimes, however, when I am with you, I can think and act like a Japanese."

The men nodded politely, and I continued.

"The same is true for you," I said. "You are all Japanese, but often you think and act like Americans when you are around me. You talk directly; you listen to my ideas. Sometimes you even laugh at my jokes."

Mr. C stared dumbfounded for a moment, then giggled nervously.

"The problem that I face is knowing when to act like you and when to act like me. Perhaps that is the secret of what you call 'internationalization.' None of us can be one color all the time. We must learn how to change naturally."

The room remained silent. I slid my legs out from under me and sat back on the floor.

"You speak from your heart," Kato-*sensei* said. "Thank you. It was a great honor having a young American like you in our office. All of us are wishing you luck, happiness, and great success. Also, we are all awaiting news in the twenty-first century reporting that Bruce-*sensei* has been elected president of the United States of America."

"*Banzai*," the teachers cheered.

.   .   .

I often wondered during my time in Japan if there was such a thing as a Japanese equivalent of the "American Dream." As I sat in this room, enjoying the comfort of this family, I thought of my early days in Sano — in the bath, the hospital, and the *kara-oke* bar. I left those initial encounters with a queasy feeling, like what a young boy feels after an interminable embrace from an over-bearing aunt. "Leave me alone," I wanted to shout. "Don't draw me so close. Let me up for air!"

After a while, however, I began to enjoy these intimate encounters — the office *enkais*, the golf practice sessions after lunch, even the banter across the open desks while I spoke on the telephone. I began to feel at one with this group, and began to admire the *amae*, or interdependence, that they felt for one another. Young Japanese people dream of earning money and owning land, just as Americans do, but my friends and col-leagues in Japan seemed most at peace when they had rekindled that sense of belonging they first knew when they were in school.

As the time came to leave, I once again offered my regrets to Mr. C and said my farewells to the men assembled around the table. I paused a final time before the green-shrouded shrine, clapped my hands together silently, and made my way toward the open door.

"Mr. Bruce," Kato-*sensei* called from behind, "please be careful of your health."

"Thank you," I said.

"*Sayonara.*"

As Cho and I stepped down from the foyer and slid into our shoes, Mr. C reached inside a cardboard box that lay open on the floor.

"Here," he said, bending down and lifting a large white envelope above his head, "our Japanese custom."

I took the envelope and walked backward out the door. As I had learned in my first days in Sano, one never exits a room frontward but backward, out of respect to the group one leaves behind.

We walked to Cho's car, and I sat down on the passenger's side. Cho sat down in the driver's seat and gestured for me to open the package from Mr. Cherry Blossom.

"It's a gift," Cho said, "for remembering his family."

I slid my hand beneath the flap on the back and lifted up the fold. I turned the envelope upside down, and out fell onto my lap a gift from Mr. C that in times of need would warm my heart and uphold my dignity: a small white towel.

A FINAL BOW

THIS BOOK BEGAN five years ago over a slice of homemade liver pie in the kitchen of Makiko Nagata. I didn't like the pie, but I'm glad I stayed for dessert.

I would like to thank the following people in Japan for their generous support and friendship: Kenji Sakuragi, Kiyoshi Tanuma, and all the men and women of the Ansoku Kyōiku Jimushō; Masashi Cho, Masami Iizuka, Toshiko Ishii, and all the teachers and students of southwest Tochigi; Dr. Shoichi Endo; Toshiaki and Shigeko Harai; Tsunemasa and Hiromi Sugiyama; Takashi and Kyoko Nagata; my colleagues Laura Sheley and Beth Myers; and my teacher Paul Scott.

Two people in particular helped turn this story into a book: my agent, Jane Dystel, believed deeply in this project; and my editor, Jane von Mehren, discovered the narrative lurking inside my manuscript. In addition, many others offered invaluable support during the writing: Gingie Halloran; Rosemary Daniell; Karen Eastman; Will Philipp; Ben Edwards; Jeffrey Shumlin; Louise Rogers; Max Stier; Cliff Johnson; Leslie Gordon; Jane Fishman; Kathleen Scott; Ben Seale; Jocelyn Ford; Beulah Harper; Henry Meyer; Ethel Mitchel; Masatoshi, Yuji, and Mie Sugiyama; and my grandmother, Aleen Feiler, who saved the letters home.

Finally, I would like to express my profound appreciation to the members of my family. My sister, Cari, has a gift for metaphor. My brother, Andrew, read every draft and improved each one with his vision. And above all I thank my parents, Jane and Ed Feiler, who first taught me how to learn: their voices echo in these pages, and this book is dedicated to them.

GLOSSARY

IT IS MUCH EASIER for native speakers of English to pronounce Japanese than the other way around. With this in mind, several tips will be helpful.

Instead of individual letters, Japanese is composed of syllables, most of which begin with a consonant and end with a vowel. When saying these words, one pronounces each syllable separately. Thus the term *kara-oke* sounds like *kah-rah OH-kay*.

Consonants are pronounced approximately as in English, with the following exceptions: *g* is always hard, as in the word *go*; *s* is said as in the word *see*; and *r* is said with the tongue touching the roof of the mouth, like the letter *d*.

The following vowels are pronounced differently in Japanese: *a* as in *father*; *e* as in *egg*; *i* as in *feet*; and *u* as in *Bruce*.

Below is a list of some of the Japanese words and phrases used in this book. The words marked with an asterisk are the ones I consider to be the main pillars of Japanese educational philosophy.

Gambate kudasai: please do your best.

aisatsu	a formal greeting
amae	a sense of dependence or attachment — like a child's feeling toward his or her mother
banzai	a cheer, literally meaning "may you live ten thousand years," formerly used only for the emperor but now acceptable for everybody

burakumin	literally, "hamlet people"; the outcast class of Japanese set up under feudal laws but officially outlawed in 1871
enkai	a drinking party
eta	literally, "people of filth"; one of the earliest names for the outcasts; derived from their association with so-called dirty professions
gaijin	literally, "person from abroad"; the most common Japanese term for a foreigner, sometimes used derogatorily
*gaman**	persevere, endure, forbear
han	a small group; used in the classroom for various activities like preparing lunch and completing homework assignments
ijime	bullying; student-on-student violence
juken jigoku	examination hell, a term used to describe the process of preparing for, worrying about, and finally taking entrance examinations for high school and university
juku	an afterschool cram school, or crammer; sometimes called *yobikō*
kanji	the name for the approximately three thousand Chinese characters that make up the core of the written Japanese language
Kansai	the "Western Plain" of Japan, centered around Kyoto and Osaka
Kanto	the "Eastern Plain," centered on Tokyo and including Tochigi Prefecture

kara-oke   literally, "empty orchestra"; a party game in which revelers sing the lyrics of a song into a microphone while a tape of an orchestra plays background music; also performed with video

*kejime**   the line of demarcation that separates two stages of life, like junior high school and high school; also, the code of behavior that one accepts with each new level of responsibility

kodomo   children; Japanese teachers commonly use this word for their teenage students

kokusaika   "internationalization"; a newly minted word to describe Japanese efforts to become more integrated with the rest of the world

kotatsu   a latticed wood frame that sits on the floor, formerly over a fire, but now over an electric heat bulb, that is used for heating the lower half of the body in winter; in summer the bulb is removed, leaving only a table

*kumi**   a homeroom class; originally a band of samurai retainers

Meiji   the period between 1868 and 1911 when Japan opened its doors to the West following a long period of isolation; named after the emperor

nanpa   the act of cruising bars, or other hangouts, and picking up dates for casual relations

onegaishimasu   a polite greeting, literally meaning "please do me the favor of . . . ," which stu-

dents utter before classes, and teachers
say at the start of the school day

san	a formal honorific attached to most names, roughly equivalent to Mr., Mrs., and Miss
*sempai/kōhai**	literally, "senior/junior"; the system of hierarchy used in schools, clubs, and companies to teach deference to seniority
sensei	teacher or master; also used as an honorific similar to *san* when speaking with educators, doctors, and professors
shinjinrui	literally, "new types"; a slightly pejorative word used to describe young people who enjoy spending money and following Western trends
Shinto	literally, "the Way of the Gods"; a set of beliefs native to Japan that celebrates the spirits living in natural things such as stones, trees, and rice
*shitsuke**	discipline; the code of behavior and human relations that the Japanese teach in schools
sushi	slices of raw fish, boiled squid, and other seafood served with sweet rice and soy sauce; not to be confused with *sashimi*, which is only slices of raw fish
tatami	a thick, woven straw mat about three feet by six feet, often edged in silk, which is used to cover the floors of traditionally appointed rooms
torii	literally, "bird perch"; a gateway, usually made with two horizontal posts and

	two crosspieces, which marks the entrance to a Shinto shrine
Tokugawa	the period between 1600 and 1868 when Japan isolated itself economically and militarily from other nations; the name comes from the family of shogun rulers who ruled during this time
uchi	one's own house, or the surroundings to which one belongs; also used to mean "I" or "we"
undō-kai	a sports festival; usually held in October around the time of National Sports Day
wabi-sabi	peaceful thoughts, peaceful actions; the feeling of tranquillity and serenity that comes from participating in traditional Japanese activities such as viewing leaves in autumn or cherry blossoms in spring

FURTHER READING

THERE ARE FOUR main types of books on Japan written by foreigners: academic books, which are written for those with an indestructible interest in Japan; business books, which seek to explain the secrets behind the Japanese economic "miracle"; survey books, in which distinguished journalists or scholars examine Japan, "its politics and its people"; and novels, the most popular of which describe a young Western man who goes to Japan and has a thwarted love affair.

The following is meant to be not an exhaustive catalogue but rather a list of some of the books I have found beneficial.

Academic Books. The best book I've read on Japanese education is *Japan's High Schools* by Thomas Rohlen (University of California Press, 1983). Other helpful works include *Society and Education in Japan* by Herbert Passim (Columbia University Press, 1965) and *The Japanese Educational Challenge* by Merry White (Free Press, 1987).

I found several studies on recent Japanese history particularly valuable: *A History of Postwar Japan* by Masataka Kosaka (Kodansha, 1982); *War Without Mercy* by John Dower (Pantheon, 1986); and *Japan's Modern Myths* by Carol Gluck (Princeton University Press, 1985).

Business Books. Bill Emmott discusses the impact of Japan's generational changes on the country's economic fortune in his book *The Sun Also Sets* (Times Books, 1990). Other important works include *Shadows of the Rising Sun* by Jared Taylor (Tuttle, 1983) and *The Reckoning* by David Halberstam (Morrow, 1986).

Kenichi Omae, a Japanese consultant, has written an excellent book on internationalization and the Japanese economy called *Beyond National Borders* (Kodansha, 1987).

Survey Books. Two notable books that seek to be comprehensive are *The Sun at Noon* by Dick Wilson (David & Charles, 1988) and *The Japanese Mind* by Robert Christopher (Ballantine, 1983).

H. Paul Varley's book *Japanese Culture* (University of Hawaii, third edition, 1984) is a fascinating overview of Japanese arts; and Roland Barthes's *Empire of Signs* (Hill & Wang, 1982) provides interesting insights into Japanese behavior.

Novels. John David Morley has written an entertaining novel about a Western man cruising the underside of Japan called *Pictures from the Water Trade* (Atlantic Monthly Press, 1985).

For a non-Western perspective, a greater number of Japanese novels are now becoming available in translation. A captivating book on the changing values of young Japanese is *A Wild Sheep Chase*, by Haruki Murakami (Kodansha, 1990). In addition, two books by the well-known author Natsume Soseki directly concern the relationship between Japanese teachers and students: *Botchan*, written in 1904, and *Kokoro*, written in 1914. Both are published in English by Charles E. Tuttle.

Finally, I would like to mention two books about teaching which were an inspiration to me: Tracy Kidder's study of a fifth-grade teacher in Massachusetts, *Among Schoolchildren* (Houghton Mifflin, 1989); and Pat Conroy's tale (Bantam, pbk., 1987) of a young man who ventures to a little-known place, gets inspired by the children he meets, and returns to write a tribute to these people and their world. The name of his book is *The Water Is Wide*, but through his words he made that distance smaller.

INDEX

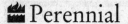 Perennial

Books by Bruce Feiler:

ABRAHAM: *A Journey to the Heart of Three Faiths*
ISBN 0-06-052509-6 (paperback) • ISBN 0-06-051863-4 (large print)
ISBN 0-06-051536-8 (unabridged audio) • ISBN 0-06-051537-6 (unabridged CD)
Feiler embarks on a quest to better understand the man at the heart of the world's three
monotheistic religions—the man all three religions call "God's Friend."
"A winning mix of insight, passion, and historical research." —*Christian Science Monitor*

WALKING THE BIBLE: *A Journey by Land Through the Five Books of Moses*
ISBN 0-380-80731-9 (paperback)
ISBN 0-694-52465-4 (audio) • ISBN 0-06-057725-8 (CD)
A fascinating, unprecedented journey—by foot, jeep, rowboat, and camel—through the
most famous stories ever told.
"An instant classic. . . . A pure joy to read." —*Washington Post Book World*

UNDER THE BIG TOP: *A Season with the Circus*
ISBN 0-06-052702-1 (paperback)
Feiler describes his season in the circus, and how a volatile mix of performers and
animals survive the daily heartaches and inevitable tragedies of life on the road.
"A colorful, sometimes unsettling pageant of circus life." —*Entertainment Weekly*

DREAMING OUT LOUD
Garth Brooks, Wynonna Judd, Wade Hayes, and the Changing Face of Nashville
ISBN 0-380-79470-5 (paperback)
The most comprehensive portrait yet painted of one of America's richest traditions:
country music—from the Grand Ole Opry to the dim light of a recording studio.
"One of the best books about country music to appear in years." —*Billboard*

LOOKING FOR CLASS: *Days and Nights at Oxford and Cambridge*
ISBN 0-06-052703-X (paperback)
With his trademark flair and humor, Bruce Feiler peeks into the privileged world of
Wordsworth and Wodehouse illuminated by his year at Oxford and Cambridge.
"Full of companionable characters, solid information, and wit." —Scott Turow

LEARNING TO BOW: *Inside the Heart of Japan*
ISBN 0-06-057720-7 (paperback)
Feiler's captivating and entertaining glimpse into the intricacies of Japanese culture,
experienced during the year he taught English in a small rural town.
"A refreshingly original look at Japan. . . . This book is a revelation."
—*Atlanta Journal-Constitution*

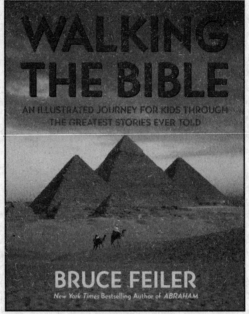